Chemical-Mechanical Polishing—Fundamentals and Challenges

MATERIALS RESEARCH SOCIETY
SYMPOSIUM PROCEEDINGS VOLUME 566

Chemical-Mechanical Polishing—Fundamentals and Challenges

Symposium held April 5–7, 1999, San Francisco, California, U.S.A.

EDITORS:

S.V. Babu
Clarkson University
Potsdam, New York, U.S.A.

S. Danyluk
Georgia Institute of Technology
Atlanta, Georgia, U.S.A.

M. Krishnan
IBM T.J. Watson Research Center
Yorktown Heights, New York, U.S.A.

M. Tsujimura
EBARA Corporation
Fujisawa-shi, Japan

Materials Research Society
Warrendale, Pennsylvania

CAMBRIDGE
UNIVERSITY PRESS

University Printing House, Cambridge CB2 8BS, United Kingdom

One Liberty Plaza, 20th Floor, New York, NY 10006, USA

477 Williamstown Road, Port Melbourne, VIC 3207, Australia

314-321, 3rd Floor, Plot 3, Splendor Forum, Jasola District Centre, New Delhi - 110025, India

79 Anson Road, #06-04/06, Singapore 079906

Cambridge University Press is part of the University of Cambridge.

It furthers the University's mission by disseminating knowledge in the pursuit of education, learning and research at the highest international levels of excellence.

www.cambridge.org
Information on this title: www.cambridge.org/9781558994737

Materials Research Society
506 Keystone Drive, Warrendale, PA 15086
http://www.mrs.org

First published 2000
First paperback edition 2013

Single article reprints from this publication are available through University Microfilms Inc., 300 North Zeeb Road, Ann Arbor, MI 48106

CODEN: MRSPDH

A catalogue record for this publication is available from the British Library

ISBN 978-1-558-99473-7 Hardback
ISBN 978-1-107-41402-0 Paperback

CONTENTS

*Invited Paper

v

*Invited Paper

PART V: <u>CMP MODELING AND FLUID FLOW</u>

PART VI: <u>PARTICLE ADHESION AND POST-POLISH CLEANING</u>

*Invited Paper

*Invited Paper

PREFACE

Chemical-mechanical planarization (CMP) has emerged over the last few years as a key enabling technology in the relentless drive of the semiconductor industry to smaller, faster, less expensive ICs. As pointed out in the National Technology Roadmap for Semiconductors, there is a need for a better understanding of the science and technology underlying CMP. Ever-increasing surface planarity and defect requirements in IC fabrication are being driven by increasing process complexity, increasing wafer sizes, and decreasing device dimensions as well as the rapid rate at which new materials are being introduced. These increasing planarity and defect requirements make the understanding of planarization fundamentals, extending to different length scales and for different material surfaces, even more important. Multi-length scale (nm to several cms) planarization and a defect-free finish of surfaces, consisting of composite regions of dielectric and metal films, can be achieved only if the characteristics like hardness and chemical functionality of the abrasive powders are tailored to the properties of the surface and to the chemical environment used during the polish process.

This symposium, "Chemical-Mechanical Polishing—Fundamentals and Challenges," held April 5–7 at the 1999 MRS Spring Meeting in San Francisco, California, was aimed at bringing together many of the active players in this field from all parts of the world. It reflects to some extent the role played both by academic institutions and multinational corporations in opening up the frontiers in the field of CMP for wider dissemination.

This volume is divided into six parts; Part I: Overview and Oxide Polishing, Part II: Pads and Related Issues, Part III: Metal Polishing—W and Al, Part IV: Copper Polishing and Related Issues, Part V: CMP Modeling and Fluid Flow, and Part VI: Particle Adhesion and Post-Polish Cleaning.

S.V. Babu
S. Danyluk
M. Krishnan
M. Tsujimura

MATERIALS RESEARCH SOCIETY SYMPOSIUM PROCEEDINGS

MATERIALS RESEARCH SOCIETY SYMPOSIUM PROCEEDINGS

Prior Materials Research Society Symposium Proceedings available by contacting Materials Research Society

Part I

Overview and Oxide Polishing

DIRECTIONS IN THE CHEMICAL MECHANICAL PLANARIZATION RESEARCH

Shyam P. Murarka
SRC Center for Advanced Interconnect Science and Technology
Rensselaer Polytechnic Institute, Troy, NY 12180

ABSTRACT

Planarized surfaces have become key to the success of advanced semiconductor devises/circuits/chips. The planarization, achieved by the use of chemical mechanical means, has enabled the interconnection of ever increasing number of devices and also the use of lower resistivity copper as the interconnect material for such devices. Chemical mechanical planarization (CMP) has now found application at several different stages of semiconductor chip fabrication and many other microelectronic applications. However, there remain a large number of nuances and effects e.g. pattern, chemical, and pad dependencies and scratching, that need to be carefully studied, evaluated and eliminated if we want to continue to progress in sub 0.1 μm (minimum feature size) regime, where the amounts of material to be removed will be small, surfaces will dominate the performance, and margin of error extremely small and unforgiving. This presentation will discuss the future needs, the CMP variables, the relationship of these variables to CMP behavior and planarity, scratch-free CMP, and size-impact on CMP outcome. A new set of goals will be presented and discussed.

1. INTRODUCTION

The planarization, achieved by the use of chemical mechanical means, has enabled the interconnection of ever increasing number of devices and also the use of lower resistivity copper as the interconnect material for such devices. The effectiveness of the chemical mechanical planarization (CMP) [1] in both improving the yield and performance of the circuits has let to its' application in the front end processes and many other microelectronic applications (in both advanced and not-so advanced devices and circuits). This tremendous growth of CMP applications is largely associated with the clever engineering developments and solutions generally credited to the tool and consumable providers. **It has led many of us to believe that there are no showstoppers.** However, as the technologies move (a) from sub-0.25μm minimum feature size (mfs) regime into the sub-0.10 μm mfs regime and multi-layered structures and (b) into the three-dimensional (3D) structures, CMP will face new challenges. In the sub-0.10μm regime of the devices, atomically flat and clean surfaces will be needed. This is true for both today's silicon based devices and new devises that have not yet been developed or found a competing edge over existing ones. For example, atomically flat and clean silicon surfaces have been oxidized to produce near-nm-thick gate oxides for the metal-oxide-semiconductor (MOS) device fabrication. [2] These MOS devices, employing polysilicon as the metal, have been shown to work with leakage currents that are a function of the atomic flatness at the Si-SiO_2 interface, increasing with increasing roughness. Note that the art of the preparation of such atomically smooth surfaces is known to mankind in both the semiconductor and optics industry. The use of the so-called chemical mechanical polishing (many a time used as acronym in place of CMP) and controlled vapor disposition techniques have been made in producing the desired atomic smoothness of the surfaces. The challenges that lie ahead refer not only to the preparation but

3

also to preserving such atomically smooth large area surfaces before, during, and after the next step in the process of fabricating semiconductor devices/circuits. In many circumstances such preparations will cover the use of large-area and multi-material (single or polycrystalline) surfaces with the unprecedented planarity requirements. In spite of the so-called no showstopper attitude, it is obvious that there remain (a) a large number of nuances and effects listed in Tables 1 & 2 and (b) a lack of fundamental understanding that need to be addressed and evaluated for the continued success of CMP into the sub-tenth micron era. Added to these challenges, and as suggested above, will be the introduction of new metals (Cu, Cu-alloys, diffusion barrier/adhesion/ promoters), new interlayer dielectrics (ILD) (polymers, aero/xerogels. others), high dielectric constant materials, and optoelectronic materials. Stringent control on the surface topography-flatness (basically, a moving target in future) will be necessary. Table 3 lists these concerns especially for the sub-tenth micron era. This paper reviews the future needs and directions in the chemical mechanical planarization research both from the National Technology Roadmap of Semiconductors (NTRS) points of view as well as the off roadmap considerations. An emphasis on the materials and chemicals aspects of the CMP will be placed in these discussions.

- Pad Related
    - generate understanding
    - role of the pad viscoelastic properties in dishing and erosion
    - optimize pad synthesis and shape for given application
    - conditioning, can we eliminate it?
- Slurry related
    - electrochemical effects
    - slurry feed and temperature
    - abrasives
    - life under load and rotation
- Feature size and loading effects
    - dishing and erosion
- Effect of film properties
    - alloying, microstructure, size
- Post-CMP cleaning
    - self-cleaning and/or passivation
- Alignment during polishing
    - tool

Table 1: Directions in CMP Research

2. MATERIALS CHALLENGES

In CMP, a pad that is made of one or more of polymer materials (sometimes with an embedded inorganic material like glass bead or an abrasive), a substrates with/without a stack of film and the final film to be polished on top, hard abrasive particles, and chemicals in the liquid slurry form the materials group that directly interact, during the process, with each other and with the materials being polished. The materials in the backing film (that supports the substrates in the

4

holder) and the platten also play an indirect role, especially when one considers the load and temperature gradients and distributions.

1. Process control
   - end point detection
   - multiple materials
   - more then one slurry
   - in-situ vs ex-situ cleaning (defectivity)

2. Temperature not exploited in a process design

3. Polished surface evaluation

4. Rotational vs. linear tool

5. Uniformity and edge exclusion
   - dishing and erosion

6. Film stack - effect on polishing

Table 2: Some of the Important Process-tool-issues

- Very small amounts of materials to be removed
  - very small polish rates, yet effective planarization (e.g. at lower pressures)

- Control the smoothness and chemical nature of the finished surface
  - that will dominate the performance and impact reliability

- Margin of error extremely small and unforgiving

- New and difficult materials, some in small amounts, others present on surface with materials with an apparent very different CMP behavior
  - materials that do not passivate or corrode, do not dissolve in aqueous media, easy to scratch or hard to fracture

Table 3: The Key Challenges (Anticipated at ≤0.1μm

2.1 PAD

Pad plays two important roles: (a) provides support against the wafer surface, allowing the wafer surface to experience the impact of mechanical and chemical forces leading to materials remove and (b) carries slurry (from the feed end to disposal) to affect planarization. In absence of the pad, which mainly provides effects related to mechanical forces, one can visualize a fast moving slurry across the wafer surface in a manner analogous to fast moving turbulent and muddy river water that polishes stones and other materials in the path of such turbulent flow of water. In absence of the slurry, the polishing will be analogous to polishing on a sand paper and the polish surface quality will depend on pad's mechanical properties relative those of the film being polished. Chemicals in the slurry help reduce the scratching on the surface because of the differential reactivity between strained and unstrained surfaces and due to the crystalline structure and atomic density on the surfaces.

Ideally, to obtain a planar surface, one should have a rigid and inert pad that can carry abrasives and chemicals. The most important problems with the ideal pad will be (a) the exacting requirements for the alignment between the pad and the sample being polished and (b) the occurrence of hydroplaning. In addition as we know pad is not a rigid material and its characteristics changes as polishing continues due to fatigue, chemical/solvent effects on pads rigidness, and wear. Similarly pad is not chemically inert material. Chemical effects cause changes in surface and possibly bulk chemistry of the pad, surface bonding between abrasive and pad, and electrochemical (charging) effects, and latter two mainly responsible for the "pad caking" or the need of "pad reconditioning". Note that pads are generally made of polymers that are spongy, porous, absorb water and lead to creation of dipoles near the surface leading to a charged state. Also the fibers (and molecule chains) on the surface experience polishing induced changes and thus cause changes in the properties of the pads.

Since pads are not rigid and inert materials, they lead to the CMP problems discussed in the previous section. A dry pad is a visco-elastic material and may have significantly higher rigidity. However, when exposed to a slurry, it becomes much less rigid as is shown in Figure 1. [3] The dynamic shear molecules measured at different frequencies decreases rapidly with soaking time in water which apparently breaks hydrogen bonds in the pad material. The effect of pad bending associated with applied load during CMP and occurring at the via or trench features (low regions) has been discussed by several researchers. [4-8]

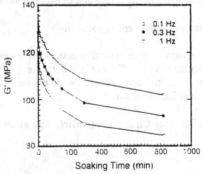

Figure 1: G' as a function of soaking time in water for a Suba IV pad (from Ref. 3)

The desire to use a rigid pad to minimize dishing and erosion and to use a softer pad to avoid hydroplaning and the need of exacting alignment requirements have led to the use dual pad (hard on top of the soft) structures. Hard material embedded pad (such as glass bead embedded ones) does not appear to provide the surface planarity and are known to produce scratching much more than dual pad systems.

2.2. FILM STACKS/UNDERLYING SUBSTRATES

When a film is polished , the underlying films and substrate may provide an effect (related to their mechanical properties, especially the compressibility) on the result of the polishing. Recent papers [9-11] have presented experimental measurements of the hardness of the film being polished and the film's polish rate in a given polishing condition, as a function of the underlying film type and thickness. Kallingal et al [9] showed that the substrate had no influence on the polish rate of the deposited SiO_2 films although it strongly influenced the nanoindentation - hardness. Tsui and Pharr measured the hardness and elastic modulus of the varying thicknesses of the aluminum film deposited on glass substrates using nanoindentation technique and different microscopic techniques and concluded that as measured properties are affected by the underlying substrate. Islam Raja and Ali [11] polished SiO_2 on Si, Al, and a film stack consisting of oxide/other dielectic film/TiN/Al all on Si in that order with oxide on top surface and Al on Si. They find "the polish rate of stacks with metal underneath is higher than that of the stack with no metal underneath." Figure 2 shows the early results of polishing Si_3N_4 deposited on film with different compressibilities. [12] Higher polishing rate for films deposited on the least compressive material agrees with the conclusions of Islam Raja and Ali. The impact of such effects on the polishing of films deposited on a surface consisting of different materials and a multilayered structure is not known and could be important in determining the planarity across the surface.

2.3. ABRASIVES

Abrasives are responsible for the mechanical abrasion (and thus removal of the material) of the surface and also for the caking (pad glazing) on the pad. Historically a variety of abrasives have been used for glass polishing and relationships between the polishing rate and characteristics to abrasive's hardness, concentration in the chemical activity, size, shape, their coagulation, and abrasive concentration in the slurry have been discussed. Note that the surfaces immersed in the liquid are generally charged by the absorption of ions from solutions. Such charges are balanced by the charge in the liquid. Chemistry, dielectric constant, temperature and surface type determine the charge type and the concentration. We have the film surface, the abrasive surface, and the pad surface that can have the charged condition. Metallic film that are not grounded or that instantaneously form a native oxide or insulating corrosion product are also charged in a liquid. If the effective charge on the abrasive has same sign as the charge on the film no polishing may occur at relatively small pressures. Such a condition also ensures polished surface to be free of the abrasive particle contamination. Similarly if the effective charge on the abrasive and on the pad have the same sign, pad conditioning may not be required since abrasives will not bond to the pad and thus pad glazing may not occur. We need to pay attention to creating such charge conditions to take advantage of the repulsive forces between identical charges.

7

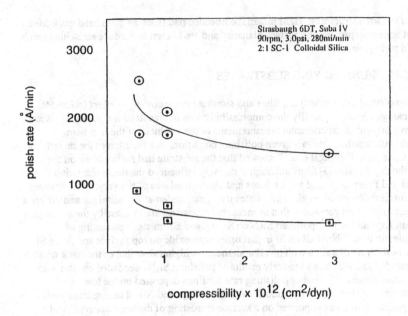

Figure 2: **Polish rate of Si₃N4 films deposited on films with different compressibilities (from Ref. 12)**

For stationary suspension of the abrasives in a liquid the potential energy can be described as a sum of the attractive and repulsive energies and determines if the particles will stay in suspension (i.e. repel each other) or will coagulate and/or form sediment. In a slurry moving across the surface of a solid, a different charging behavior occurs due to the continuous stripping of the charges at the liquid/solid interface. The charges on the solid surface and in the liquid near this surface are now separated by a hypothetical shear plane and an electrokinetic potential (commonly known as Zeta potential) develops at this shear plane. The Zeta potential is a function of the surfaces involved, electrolyte type, concentration and pH.

Generally the practice of CMP, in the Silicon IC fabrications, has been empirical or at best based on experiences of the silicon and glass polishing. It is obvious from this discussion on the abrasives that investigations of the charge state and its manipulations (by chemistry) in optimizing the CMP process are needed as the next generation optimizations become necessary for the silicon integrated circuits.

Finally, it is important to mention the role of lubrication hydrodynamics in CMP. [13-14] Although it is known that the slurry forms a thin lubricating film between the wafer and the pad [15], most of the models do not incorporate the effects of the slurry flow and the chemical reactions which affect CMP. To the author's knowledge, the first published work which presents a wafer-scale slurry flow analysis is by Runnels and Eyman. [16] They solve the steady-state three-dimensional Navier-Stokes equation numerically using a finite-element scheme in the region between the wafer and the pad. Then the process is iterated to obtain a stable position of the wafer which balances the moment about the gimbaling point. Considerable simplifications for solving the slurry flow may be achieved if one recognizes that the reduced Reynolds number is very small and therefore the lubrication approximations used in the theory of slider bearings are valid.

3. CHEMICAL ASPECTS OF CMP

Preston's equation for polishing rate's relation with pressure and relative linear velocity does not include any chemical factors. Chemistry, however, plays a very important role in determining the CMP process and its outcome including the level and type of contamination on a finished product. Although a large number of papers have referred to this important role of chemistry, there has been no sound fundamental understanding created to challenge the validity of the Preston's equation or its verifications of. [17] Most of the slurry formulations are intuitively created and experimentally optimized.

Slurries in general are made with water except in few cases where organics (to control viscosity, or as surfactants, or as surface passivators) have been used to optimize the CMP process. [18,19] Chemicals are added to affect dissolution of the abraded material, to keep abrasives in suspension as discussed above, and to affect a sacrificial layer on the polishing surface. The growth of the sacrificial layer is self-limiting so that as polishing continues, abrasion removes this layer only and does not impact underlying film (e.g. WO_x layer on W, or hydrated SiO_2 layer on SiO_2). The latter effect of the chemicals has led to the concept of so-called scratch-free CMP. [20-21] The slurry chemicals also affect the mechanical properties of the abrasives (besides their suspension behavior) and pads as discussed earlier.

Effects of chemistry are best illustrated through electrochemical measurements and principles, the focus of optimizing slurry development programs in industry and academia. For example, nitric acid dissolves copper by forming $Cu(NO_3)_2$ which has high solubility in water, this does not allow an oxide formation on surface but dissolves abraded copper readily. NH_4OH on the other hand forms hydroxide with copper. Copper hydroxide has limited solubility in water and can be useful in invoking the concepts of using a continuously growing (at a steady state and self limiting) sacrificial layer to minimize scratching. However, abraded material may not readily dissolve except that, in this case, NH_3, in NH_4OH solutions, complexes with Cu to form $[Cu(NH_3)_x]^{2+}$ or $[Cu(NH_3)_x(H_2O)_{6-x}]^{2+}$ with very high solubility in water. Complexing lowers the activity of Cu^{2+} ions and thus allows the presence of more ions in solution, effectively increasing solubility. The complexing behavior of Cu^{2+} and NH_3 and its role in CMP of copper has been investigated by the use of the electrochemical potential measurements in-situ (or ex-situ). [22] These studies clearly establish a range of pH and concentrations necessary for the effective complexing and dissolution. Figure 3, for example, shows a plot of the polish rate vs. change in the measured potentials with various ammonium salts (added to deliver the same total NH_3 weight percent concentrations). [22] According to the pH-potential (Pourbaix) diagram for the Cu-NH_3-H_2O system, complexing of Cu^{2+} with NH_3 is most likely only in NH_4OH slurries (not favored in NH_4NO_3 or NH_4Cl slurries that are on acidic pH side of the pH- potential diagram. [22] Thus in the NH_4NO_3 slurry most of copper remains uncomplexed, yielding an overall lower polish rate. Perhaps no dissolution (by complexing mechanism) occurs in the NH_4Cl slurry. It has also been shown that the addition of increasing amounts of $Cu(NO_3)_2$ to a NH_4OH slurry lowers the pH and reduces the availability of NH_3 for complexing leading to lowering of the polish rate.[1]

The use of surfactants can be judiciously made to polish materials that do not form corrosive products. Neirynck et al [18] showed that small additions of a surfactant to a slurry significantly increased the polish rate of benzocyclobutene (BCB). Additions of glycerol to a copper slurry caused the polish rater of copper initially to increase with glycerol concentration and then to decrease with continued increase in glycerol concentration, suggesting a role of increasing viscosity and charge-interaction forces that are affected by the presence of glycerol. [19]

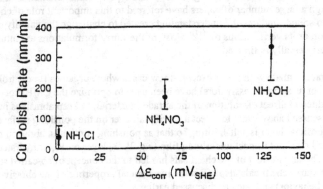

Figure 3: Polish rate vs. the corrosion potential change measured in situ during onset and end of the CMP process (Cu in alumina slurries containing NH_4OH, NH_4NO_3, or NH_4C)

4. CONCLUSIONS

This paper briefly reviews the present knowledge and raises issues that need to be addressed in the future if we want to continue using CMP for the removal of smallest amounts of material on top of features that will lie in sub-100nm range. An understanding of various variables on dishing on erosion and using this understanding to eliminate them are needed for all advanced applications. In this researcher's viewpoint mechanical effects, although will remain very important, will be dominated by chemical effects that should be used for precision in removing nanoscale materials.

5. ACKNOWLEDGEMENTS

Author will like to thank Professors David Duquette, Ronald Guttman, William Gill, and Minoru Tomozawa, Drs. Joseph Steigerwald, Weiden Li, and Jan Neirynck, and his students for their insights and contributions to author's knowledge base. He will also like to thank SRC for the support of his research activities.

REFERENCES

1. "Chemical Mechanical Planarization of Microelectronic Materials," J. M. Steigerwald, S. P. Murarka, and R. J. Gutmann (Wiley-Interscience, NY, 1997).
2. S. J. Hillenius, Lucent Technoogies, Murray Hill, NJ, Private Communication.
3. W. Li, D. W. Shin, M. Tomozawa, and S. P. Murarka, Thin Solid Films 270, 601 (1995).
4. S. R. Runnels, J. Electochem Soc. 141, 1900 (1996).
5. W. Li, D. W. Shin, M. Tomozawa, and S. P. Murarka, Thin Solid Films 270, 601 (1995).
6. J. M. Steigerwald, R. Zirpoli, S. P. Murarka, D. P. Price, and R. J. Gutmann, in Proc. Adv. Metallization Conf. at Austin, TX Oct. 3-5 (1994), (MRS, Pittsburgh, 1995), p. 173.
7. K. Achutlan, J. Curry, M. Lacy, D. Campbell, and S. V. Babu, J. Electronic Materials 25, 1628 (1996).
8. S. C. Runnels, J. Electronic Materials 25, 1574 (1996).
9. C. G. Kallingal, M. Tomozawa, and S. P. Murarka, J. Electrochem. Soc. 145, 1790 (1998).
10. T. Y. Tsui and G. M. Pharr, J. Mater-Res. 14, 292 (1999).
11. M. Islam Raja and I. Ali, in the Proc. of 1995 VMIC Conference, Santa Clara, CA (Univ. of South Florida, Tampa, FL, 1995), p. 453.
12. Y. -Z. Hu, R. J. Gutmann, and S. P. Murarka, Unpublished work (1996).
13. S. Sundararajan, D. G. Thakurta, D. W. Schwendeman, S. P. Murarka, and W. N. Gill, J. Electrochem. Soc. 146, 761 (1999).
14. J. Ticky, J. Levert, L. Shan, and S. Danyluk, J. Electrochem. Soc. 146, (1999).
15. T. Nakamura, K. Akamatsu, and N. Arakawa, Bull. Japan Soc. Prec. Engg. 19 120 (1985).
16. S. R. Runnels and L. M. Eyman, J. Electrochem. Soc. 141 1698 (1994).
17. B. Zhao et al, paper presented at the 1st IITC San Francisco, June 1998.
18. J. M. Neirynck, G. -R. Yang, S. P. Murarka, and R. J. Gutmann, Thin Solid Films 290-291, 447 (1996).
19. D. Permana, S. P. Murarka, M. G. Lee, and S. I. Beilin, Electrochem. Soc. Proc. 96-22, 206 (1997), Electrochem. Soc. Pennington, NJ.
20. "Snatch-free CMP of Electronic Materials at Rensselaer - a Science-to-Engineering Approach," S. P. Murarka, presentation at the Solid State Technology Chemical Mechanical Planarization Seminar, March 31, 1997, San Jose, CA.
21. S. P. Murarka, Mat. Rs. Soc. Symp. Proc. 511, 277 (1998).
22. J.M. Steigerwald, D. J. Duquette, S. P. Murarka, and R. J. Gutmann, J. Electrochem. Soc. 142, 2379 (1995).

THE INFLUENCE OF pH AND TEMPERATURE ON POLISH RATES AND SELECTIVITY OF SILICON DIOXIDE AND NITRIDE FILMS

W. G. America*, R. Srinivasan**, S.V. Babu**
*Microelectronics Technology Division, Eastman Kodak Company, Rochester, NY 14650-2024
**Center for Advanced Material Processing, Clarkson University, Potsdam, NY 13699-5705

ABSTRACT

Planarization of dielectric films, silicon dioxide, silicon nitride, etc. by Chemical-Mechanical Polishing (CMP) is regulated and moderated by the interaction of the abrasive particles and chemicals in solution with the film surface through complex chemical and physical processes. Changes in the slurry properties have a profound effect on the polishing chemistry and relative removal rates of dielectric films. Common slurry properties include pH, temperature, abrasive particle composition, its size and shape, degree of agglomeration, and weight percent, and chemical composition. While the slurry vendor has control over most slurry properties, the pH and temperature can be controlled during the polishing process by the user and can have a strong influence. Data are presented highlighting the influence of pH and temperature on the CMP of both blanket and patterned silicon dioxide and silicon nitride films.

INTRODUCTION

CMP relies on mechanical abrasion coupled with chemical activity to remove and ultimately planarize the top film or films on wafers during semiconductor processing. The mechanical variables during CMP, including table speed, down force, pad hardness, etc., are typically used to control rate, planarity, and uniformity. The chemical part of CMP is usually exploited to achieve selectivity and help address issues of pattern dependence such as erosion and dishing. Polishing of dielectric films like silicon dioxide and silicon nitride is usually carried out using slurries containing silica or ceria abrasive particles. The silica slurries are typically prepared with pH greater than 8, while ceria slurries have pH ranging from 4 to 10. In both cases, the pH of the suspension can have a strong effect on the polish rate. Knowledge of the effect of pH on polish rate and film selectivity can be useful in understanding and controlling the CMP process.

EXPERIMENTAL

Polishing experiments were carried out using two types of pads and two CMP tools. Grooved and stacked IC1400/Suba IV (Rodel) pads in a single spindle, manual Strasbaugh polisher were used in experiments using ceria slurries. Perforated and stacked IC1000/Suba IV (Rodel) pads in a Strasbaugh Model 6DS-SP polisher were used in experiments using silica slurries. Unless mentioned otherwise, the temperature of the slurry was maintained at 25 °C for ceria slurries and 30 °C for silica slurries on the polishing tables. The films studied were PECVD TEOS silicon dioxide and LPCVD silicon nitride on 150 mm diameter silicon wafers. Film thicknesses, before and after CMP, were measured optically with either an ellipsometer or interferometer.

The effect of pH on polish rate was explored for both ceria and silica slurries. A 5 wt % ceria slurry (Ferro Electronics Division, Ferro Corporation) was used for all the ceria experiments. Two types of ceria particles, one with an average diameter of 440 nm and an isoelectric point (IEP) of 8.5 and another with an average diameter of 140 nm and an IEP of 6.5

13

were used in the polishing experiments. The silica slurry used was SS-25 (Cabot Corporation) diluted 1:1 with DI water. Adjustments of the slurry pH was carried out with either 11.6 N HCl or 40% KOH solution. Ceria polishing was carried out with a 30 rpm table and spindle speed, a 240 mL/min slurry flow rate and a 5.6 psi down force. Silica polishing was carried out with a table and spindle speed of 75 rpm each, 200 mL/min slurry flow rate, and with a 6 psi down force and a 1.5 psi wafer back pressure. The effect of the temperature of ceria slurry on polishing rate was investigated using heated slurry and measuring the temperature of the slurry as it flowed onto the polishing table. Warm water at the temperature of the slurry was flowed over the pad for three minutes before each run to ensure that the pad does not reduce the temperature of the slurry appreciably.

The effect of pH on patterned TEOS oxide wafers was also investigated using the MIT CMP Characterization Mask Set pitch mask [1]. The pitch mask is a 6 x 6 array of various equal width lines and spaces. The lines and spaces have widths ranging from 2 μm to 1000 μm. The 12 mm pitch pattern was first created in photoresist across the entire wafer. The exposed oxide was plasma etched to a depth of about 7000 Å below the surface. This created an array of 36 sub-arrays with equal width lines and spaces so that half the area was lines and the other half spaces. The wafers were polished with silica slurry at 3 different pH values. The oxide thickness at the plateau area (starting surface of the oxide) for 10 sub-patterns of line widths 40, 60, 80, 100, 125, 150, 180, 200, 250, and 500 μm were measured before and after CMP. The measurements were made in each sub-array in increasing order of line/space width. Five dies were measured along the diameter of the wafer.

RESULTS AND DISCUSSION

Effect of Temperature on Polish Rate for Ceria Slurry

The effect of temperature of slurry containing ceria particles (average diameter 440 nm and IEP 8.5), on the removal rates of TEOS oxide and silicon nitride was found to be weak as seen in Figure 1. This is similar to the effect of temperature of a silica based slurry on silica polishing in the temperature range of 20-50 °C for the slurry with the pad maintained between 25-30 °C [2]. The pad hardness decreases with an increase in temperature [2] and the associated reduction in removal rate may have compensated for any increased chemical removal caused by the increased temperature.

Effect of pH on Polish Rate for Ceria and Silica Slurries

Figure 2. shows the change of polish rate of silicon dioxide and silicon nitride as the pH of the silica slurry changes from 9.7 to 11.3. There is a steady increase in the TEOS polish rate as the pH increases which is expected since silicon dioxide is known to have an increasing dissolution rate with increasing pH. The silicon nitride, on the other hand, shows a drop in polish rate as the pH increases, changing from about 700 Å/min to just over 500 Å/min at pH of 11.3. The solubility of silica increases substantially with pH in the pH range of 9-11 [3] and the possible decrease in silica particle size with increasing pH could result in reduced nitride polish rate.

The effect of the pH of slurry containing ceria particles (average diameter 440 nm and IEP 8.5), on the removal rates of TEOS oxide and silicon nitride was found to be weak, Figure 3. Polishing with a different ceria slurry (average particle diameter 140 nm and IEP 6.5) also shows a weak pH dependence for TEOS oxide removal rate but the silicon nitride polish rate decreases with pH, as seen in Figure 4. The dependence of the polish rate for both the ceria slurries on pH

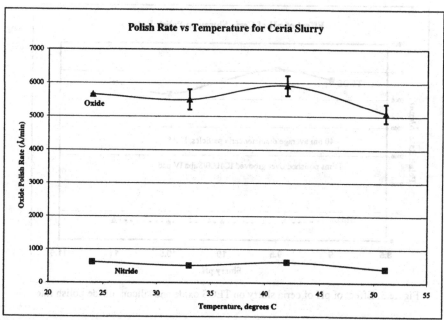

Figure 1. Variation of polish rate with temperature for TEOS oxide and silicon nitride. The slurry was 5 wt % ceria, 440 nm diameter, IEP 8.5 with pH adjusted to 10.

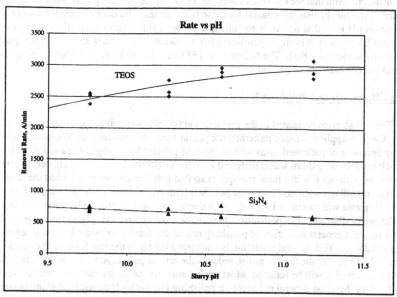

Figure 2. Variation of polishing rate of TEOS oxide and silicon nitride with pH for SS-25 (1:1) using an IC1000/SubaIV stacked pad.

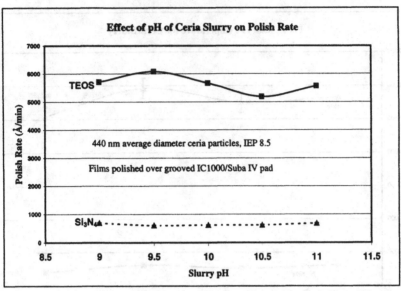

Figure 3. Effect of pH of ceria slurry on TEOS oxide and silicon nitride polish rate.

is quite different from that for the silica slurry for TEOS oxide polishing indicating a possible difference in the mechanism of removal for these two slurries. The dependence of silicon nitride polish rate on pH is similar for silica and one type of ceria slurry (average particle diameter of 140 nm and IEP of 6.5). It is slightly different from another type of ceria slurry (average particle diameter 440 nm and IEP 8.5). The behavior of TEOS oxide and silicon nitride polishing rate with ceria slurry continues to be studied.

Pattern Effects, pH and Polishing Rates

The polish rates presented in the previous paragraphs were obtained using blanket films. Usually, CMP is applied to make patterned nonplanar films flat and smooth. We have studied the polishing behavior of patterned films with changes of pH and for a range of feature widths. The MIT CMP pitch mask pattern was transferred into the TEOS oxide film. This created an array of 36 sub-arrays with equal width lines and spaces so that half the area were lines and the other half spaces. The resulting patterned films were polished with silica slurry in the pH range of 11 to 9. The polishing rate was calculated for each pH value at the same line/space location on five dies across the wafer diameter. The polishing rate increased with increasing pH, in the range of 9.3-11. We also see a general reduction of polishing rate as the line/space width increases, except for the data at 125 nm. This general reduction is not surprising since the rate follows Preston's equation [5], increasing the area of polish reduces the rate of polish. Also, for wider lines, the effect of edge erosion will be reduced. Most importantly, the polishing rate for each line/space pattern follows the same general reduction in polishing rate with pH as with polishing planar TEOS films with a silica slurry (see Figure 2).

16

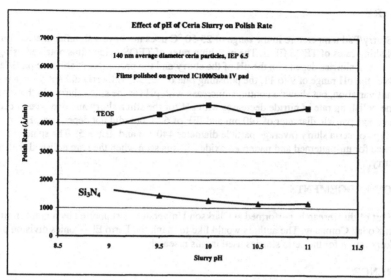

Figure 4. Effect of pH of ceria slurry on TEOS oxide and silicon nitride polish rate.

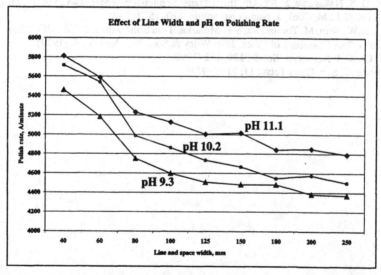

Figure 5. Polishing rate of TEOS for sub-arrays of varying width of line and spaces and the effect of slurry pH.

SUMMARY

Slurry/Pad temperature in the range of 25-50 °C does not have a significant effect on ceria polishing rates of TEOS films. The polishing rates of TEOS oxide films obtained using silica and ceria slurries depend on the pH of the slurry and the properties of abrasive particles used. Over the pH range of 9 to 11, the polishing rate of TEOS with ceria did not show any significant variation, but shows a continual increase with pH for the silica slurry. In the same pH range, the polishing rate of nitride decreases with pH for the silica slurry and one type of ceria slurry (average particle diameter of 140 nm and IEP of 6.5), but did not depend on pH for another type of ceria slurry (average particle diameter 440 nm and IEP 8.5). The same pH dependence for un-patterned and patterned oxide films is seen when they are polished with a silica slurry.

ACKNOWLEDGEMENTS

Part of this research, performed at Clarkson University, is supported by a grant from Eastman Kodak Company. The authors would like to thank the Ferro Electronics division of Ferro Corporation for the ceria slurries used in this research.

REFERENCES

[1] B. E. Stein, D. O. Ouma, R. R. Divecha, D. S. Boning, J. E. Chung, D. L. Hetherington, C. R. Harwood, O. S. Nakagawa, S. –Y. Oh, IEEE Trans. Semiconduct. Manufact., vol. 11, pp 129-140, Feb. 19982.L. M. Cook, J. Non-cryst. Solids, 120, 152 (1990)
[2] W. Li, D.W. Shin, M. Tomozawa, S.P. Murarka, Thin Solid Films, 2(70), 601 (1995)
[3] R. K. Iler, The Chemistry of Silica, John Wiley & Sons Inc., New York (1979)
[4] L. M. Cook, J. Non-cryst. Solids, 120, 152 (1990)
[5] F. Preston, J. Soc. Glass Tech., 11, 214 (1927)

THE STUDY OF OXIDE PLANARIZATION USING A GRINDSTONE

Hiroyuki Yano, Katsuya Okumura, Fumito Shoji,
Yutaka Wada*, Hirokuni Hiyama*, Norio Kimura*

Microelectronics Engineering Laboratory, TOSHIBA CORPORATION,   Yokohama, Japan;
*CMP Division, EBARA CORPORATION,   Fujisawa, Japan

ABSTRACT

Oxide planarization using a specially fabricated "grindstone" was investigated. Using the grindstone, better planarity compared with the conventional CMP technique was demonstrated. Interestingly, it was also found that the oxide removal rate became very slow after the oxide surface became planar. This self-stop was thought to be influenced by the isolated abrasives from the grindstone. The dependence on the tool structure and the conditioning was investigated to prove the model. Also the defect issue is presented in this paper.

INTRODUCTION

As a good planarization method, CMP has been introduced in the semiconductor manufacturing. But still the CMP planarization could not achieve the global planarization. To achieve the global planarization, one approach is the development of the slurry and another is the improvement of the pad. By the slurry development, the global planarization was demonstrated using the unique pressure sensitive slurry. [1,2] On the other hand, harder pad is suitable for better planarity but sometimes harder pad has the poor ability to hold the abrasives and CMP removal rate is slow. So the possibility of a grindstone for the oxide planarization was investigated. A grindstone has two special characteristics. One is high compressivity that is desirable for good planarity. Another is the fixed abrasive that is good for high removal rate. So it looks reasonable to apply a grindstone for the oxide planarization.

EXPERIMENTAL

Fig.1 shows the schematic of the grindstone used in this experiment. The grindstone is consist of Ceria and polyimide. In the grinding process, de-ionized water is supplied to the grindstone at the flow rate of 300ml/min. As a reference, double layered hard type polyurethane pad is used. This case, 10% solid content silica slurry that pH is adjusted to about 10 by KOH is supplied to the pad at the flow rate of 200ml/min. The carrier pressure is $400gf/cm^2$ and the carrier and the table speed is 30rpm for both case. The pad conditioning was performed by the conventional Diamond(100grit)/Nickel conditioning plate.

The sample was prepared by the following procedure. 700nm deep trenches with various size were created on 8 inch silicon wafer. 1400nm of SiO_2 film is deposited on this wafer by plasma CVD with TEOS source.

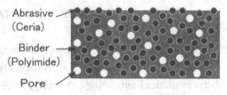

Fig. 1 Schematic of grindstone

Following the grinding or CMP process, the oxide thickness was measured by the conventional measurement tool using optical interference and the planarity was characterized by the profilometer. For the defect analysis, the defect inspection tool based on light scattering and AFM was used.

RESULTS AND DISCUSSION

Planarity

Fig.2 shows the characteristics of planarity depending on oxide removal on upper area. The space width is 1000 μ m in this case. The grinding shows the excellent planarization and almost same as the ideal curve. But CMP shows poor planarization and 1000A of step remains after 700nm of oxide removal. Fig.3 (a) and Fig.3 (b) show the profile of the oxide surface. CMP case, not only the oxide on upper area but also the oxide on lower area is removed after the step becomes small. Because of the pad compressivity, the pad surface reached the

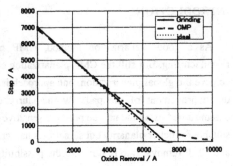

Fig.2 Comparison of CMP planarity with grinding

bottom of the groove. So the oxide in the groove is also removed and the corner of the groove is rounded after CMP. On the other hand, the oxide on lower area is not removed in grinding case. Because the grindstone is hard enough and the grindstone does not reach the bottom of the groove. So the oxide on lower area is not removed and the corner still remains after the grinding. The difference of planarization capability is thought to be the difference of the compressivity of the grindstone and CMP pad.

Self-stop

To investigate the planarization more detail, the oxide thickness depending on the grinding time and CMP time was examined in Fig.4. CMP case, not only the oxide on upper area but also the oxide on lower area is removed after the step becomes small but the oxide on lower area is not removed in grinding case as mentioned in above.

Also Fig.4 shows the special characteristic of the grinding. After the oxide surface becomes planar, oxide removal rate becomes very slow. It looks as if the removal is self-stopped after the planarization. The oxide removal rate on the blanket wafer and on the patterned wafer was compared in Table 1 to understand the reason. The CMP oxide removal rate is about 1200A/min on the blanket wafer and about 2000A/min on the patterned wafer in the beginning of polishing. CMP rate on the patterned wafer is about 1.7 times higher compare to the blanket wafer and it is reasonable because the pattern density is about 60%. On the other hand, the grinding rate is about 200A/min on the blanket wafer and about 3500A/min on the patterned wafer in the beginning of grinding. The grinding rate on the patterned wafer is about 17 times higher compare to the blanket wafer and it could not be explained from the pattern density.

So to explain this self-stop, we suggested the model and some experiment was carried out in the following.

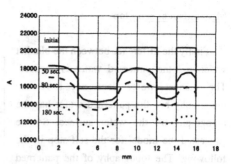

Fig.3 (a)　Profile of CMP planarization

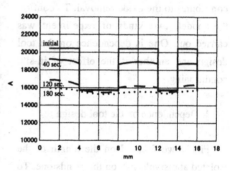

Fig.3 (b)　Profile of grinding planarization

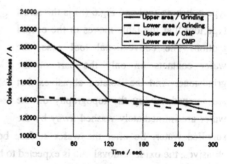

Fig.4　Comparison of CMP planarization with grinding

21

Mechanism of self-stop

Firstly the difference between the blanket wafer and the patterned wafer is focused. The blanket wafer does not have the topography but the patterned wafer has the topography on the wafer surface. So we suggested the model for the self-stop in the following. The topography of the patterned wafer could scrape out the Ceria abrasives from the grindstone as shown in Fig.5. And this scraped out or isolated abrasive mainly contributes to the oxide removal. To confirm the model, two kinds of experiment was carried out. One is dependence on the tool design and another is the effect of in-situ conditioning.

1. Dependence on the tool design

Firstly we focused on the length of the isolated abrasive's stay on the grindstone. To change this, two kinds of tool was used and compared the grinding characteristics. Fig.6 (a) is the schematic of the conventional table type polishing tool and Fig.6 (b) is the scroll type tool that the table moves in the orbital motion. The conventional tool case, the wafer contacts on only the part of the grindstone. And the isolated abrasives that are scraped out by the topography on the wafer are immediately washed away by the supplied water. So after the wafer surface becomes planar and stops to scrape out the abrasives, the oxide removal rate is expected to become very slow. On the other hand, most of grindstone surface is covered with the wafer in the scroll type case. The scraped out isolated abrasives are trapped between the wafer and the grindstone. So even after the wafer surface becomes planar and stops to scrape out the abrasives, these trapped abrasives are expected to

Table 1

| | Oxide removal rate [A/min] | | ratio |
	Blanket wafer	Patterned wafer	
CMP	1200	~2000	~1.7
Grinding	<200	3500	>17.5 (self-stop)

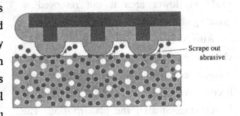

Fig.5 Schematics of scraping out abrasive

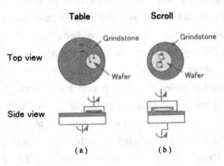

Fig.6 Schematic of tool designs

continue to remove the oxide film. Fig.7 show the oxide thickness depending on the grinding time for these two kinds of tools. The conventional tool case, the oxide removal stops after the oxide surface becomes planar as mentioned in the previous. The scroll type tool case, the oxide removal rate is high and the oxide removal does not stop even after the wafer surface becomes planar as expected. Also the table 2 shows the oxide removal rate on the blanket wafer and the patterned wafer for both tools.

On the table type tool, the oxide removal rate is very slow before and after processing on the patterned wafer. But on the scroll type tool, the removal rate is slow before processing the patterned wafer and high after processing the patterned wafer. Because the isolated abrasives are not generated before the patterned wafer process and slow

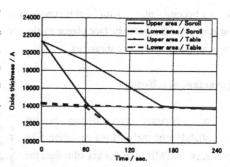

Fig.7　Comparison of planarization with scroll type and table type

Table 2

	Oxide removal [A/min]		
	Blanket (before patterned)	Patterned	Blanket (after patterned)
Table	< 200	3500	< 200
Scroll	< 500	7000	4000

removal rate. And the isolated abrasives are scraped out by the topography of the patterned wafer and enough isolated abrasives still remained on the grindstone and higher removal rate after processing the patterned wafer. So the different characteristics of two kinds of tools can explain the generation mechanism and the importance of the isolated abrasives.

2. Effect of in-situ conditioning

Secondly we focused on the scraping out of the isolated abrasive. To scrape out the abrasives intentionally from the grindstone, in-situ conditioning was tried. The comparison of oxide thickness depending on the grinding time with and without in-situ conditioning is shown in Fig.8. Without the conditioning, the oxide removal stops after the wafer surface becomes planar. But the grinding does not stop even after the wafer

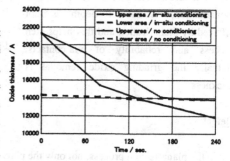

Fig.8　Comparison of planarization depending on conditioning

surface becomes planar with in-situ conditioning. Because the isolated abrasives are scraped out even after the wafer surface planar by the conditioner. So the self-stop can be explained by the generation of isolated abrasives.

Advantage of self-stop

The advantage of self-stop is the controllability of the post-planarization film thickness. In CMP case, thicker film than the initial step needs to be removed because of poor ability of CMP planarization. But the oxide removal rate has the within wafer and wafer to wafer non-uniformity. And these non-uniformity affects the post-CMP film thickness variation. On the other hand, the grinding has excellent planarization capability and self-stop phenomenon. The self-stop only remove the oxide on upper area and automatically stops oxide removal after the wafer surface becomes planar. So the grinding does not cause the oxide thickness variation and only the deposition cause the oxide thickness variation as shown in Fig.9. This controllability of post-planarization film thickness is very helpful for the semiconductor manufacturing. In ILD planarization, via-contact depth is not affected by planarization and this helps etch process and reliability of via-contacts. Actually the grinding oxide removal is thickness

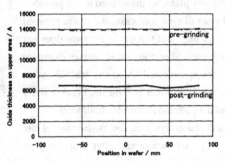

Fig.9   pre-grinding and post-grinding oxide thickness

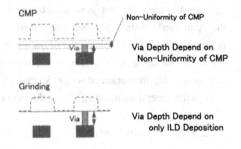

Fig.10   Schematic of oxide planarization by CMP and grinding

Defect

In the planarization process, not only the planarity but also the defect is a big issue. So the number and the size of defects are analyzed. Table 3 shows the number of defects after CMP and grinding. For both case, most of defects was scratch. This shows that the conventional sponge cleaning is applicable as post-grinding cleaning. And the size of scratch was analyzed by AFM. From the AFM analysis, the typical size of the scratch was 2.5 μ m wide and 6nm deep. The number and the size of scratch was less than our expectation. The reason would come from the hardness of Ceria abrasives.

The hardness of Ceria is almost same as the oxide. So the Ceria would not make large scratches on the oxide. So rather than Ceria abrasives, also the foreign material needs to be controlled in the grinding process.

Table 3

after C M P	after Grinding
2 4	5 5

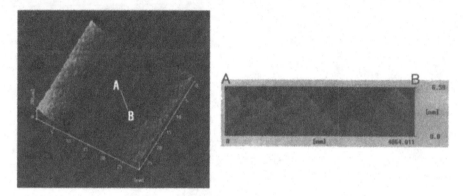

Fig.11   Scratch analyzed by AFM

CONCLUSION

The oxide planarization using the grindstone achieved the excellent planarization. This came from the higher compressivity of the grindstone compare to the CMP pad. Also the interesting phenomenon that the oxide removal rate became very slow after the oxide surface became planar was found. From the investigation, this self-stop was influenced by the isolated abrasives scraped out from the grindstone. This self-stop is very useful phenomenon for the manufacturing. Because the oxide removal stops after planarization and non-uniformity of oxide removal does not affect the post-planarization oxide thickness. The case the grinding is used for ILD planarization, via-contact depth variation only depend on the non-uniformity of the oxide deposition. Also the oxide removal can be minimized and only need to deposit the minimum of the oxide film because of the excellent planarization of grinding. This is good for the productivity. From the defect analysis, the defect density and size was comparable to CMP planarization. So the grinding can be thought as a good alternative for the oxide planarization.

References

H. Nojo et al., proc. IEDM, (1996), 349

Y. Shimooka et al., proc. VMIC, (1997), 119

ABRASIVE EFFECTS IN OXIDE CHEMICAL MECHANICAL POLISHING

UDAY MAHAJAN, MARC BIELMANN AND RAJIV K. SINGH
Department of Materials Science and Engineering and Engineering Research Center for Particle Science and Technology, University of Florida, Gainesville, FL 32611

ABSTRACT

In this study, we have characterized the effects of abrasive properties, primarily particle size, on the Chemical Mechanical Polishing (CMP) of oxide films. Sol-gel silica particles with very narrow size distributions were used for preparing the polishing slurries. The results indicate that as particle size increases, there is a transition in the mechanism of material removal from a surface area based mechanism to an indentation-based mechanism. In addition, the surface morphology of the polished samples was characterized, with the results showing that particles larger than 0.5 μm are detrimental to the quality of the SiO_2 surface.

INTRODUCTION

Chemical Mechanical Polishing (CMP) has become the industry-wide standard for achieving global planarization of metal and dielectric films in Multilevel Metallization schemes [1]. Oxide CMP, in particular, has been integrated most successfully into the manufacturing environment. However, the fundamental processes underlying the polishing process have yet to be fully understood. Most of our knowledge in this regard has been derived from previous studies on glass polishing [2], [3]. Those results may not be applicable under the conditions of present-day CMP, which utilizes extremely small (submicron sized) abrasives to polish thin metal and oxide films. In addition, the effect of abrasive size on polishing rate and surface smoothness has not been fully understood. Different researchers have obtained different (and often contradictory) conclusions as to how particle size influences polishing properties of a slurry. Jairath et al. [4] observed that polish rate increased with both particle size and concentration. Xie and Bhushan [5] obtained similar results for polishing of copper and ferrite with diamond and alumina abrasives. However, Cook [2] and Sivaram [6] proposed that polish rate is independent of abrasive size, and Izumitani [3] observed that decreased abrasive size led to higher polish rates for optical glasses. Recent studies conducted by us [7] on particle size effects in Tungsten CMP have shown that polish rate increases as particle size is reduced.

In this paper, the results of a systematic study on the effect of particle size on polish rate and surface roughness in silica CMP will be presented.

EXPERIMENT

Sol-gel silica particles of four different sizes, viz. 0.2 μm, 0.5 μm, 1.0 μm and 1.5 μm were obtained from Geltech® Corporation. Particle size analysis on the slurries was carried out using a Honeywell Microtrac® UPA 150 particle size analyzer, which utilizes the dynamic light scattering technique. In addition, Transmission Electron Microscopy (TEM) and Scanning Electron Microscopy (SEM) was also used for determining particle size and shape.

Polishing experiments were carried out p-type Silicon wafers on which a 1.5 μm thick SiO_2 layer was deposited by PECVD. Slurries were prepared by dispersing the silica particles in DI

27

water, using an ultrasonic probe to break up agglomerates. The pH of all the slurries was adjusted to 10.50 by using NaOH and HCl. The polishing was carried out on a Struers Rotopol 31 polisher, using IC 1000/Suba IV stacked pads (supplied by Rodel Inc.), a pressure of 7.0 PSI, and a rotation speed of 150 rpm for both the pad and the wafer. Polishing rates were determined by measuring film thicknesses of the samples before and after the polishing experiments, using ellipsometry. Atomic Force Microscopy (AFM) was used to characterize the surface roughness and morphology of the samples.

RESULTS AND DISCUSSION

Electron micrographs of the particles are shown in Fig. 1. It can be seen that the particles are extremely spherical and almost monosized. These results were also confirmed by particle size analysis, the results of which are shown in Fig. 2. Fig. 3 shows the oxide removal rate as a function of solids loading for the different particle size slurries. As can be seen from the figure, the 0.2 µm particles show an increasing removal rate with solids loading, which flattens out after 10wt.%. On the other hand, the 0.5 µm particles first show an increase in removal rate from 2 to 5%, followed by a drop. The 1.0 µm and 1.5 µm particles show very high polish rates for low solids loading, which then decrease steadily as the solids loading is increased.

The above results seem to indicate a transition from one removal mechanism to another as particle size is increased. In another paper [8], we have described two mechanisms for polishing, an indentation-based mechanism and a surface area based mechanism. The mathematical expressions for particle size and solids loading dependence have been derived from an expression for penetration depth of an abrasive particle, which was developed by Brown et al. [9]. According to the surface area based model, polishing rate depends on the total contact area between the abrasive particles and the surface being polished. The contact area as a function of particle size and abrasive concentration is given by the following expression:

$$A \propto C_0^{1/3} \cdot \phi^{-1/3}$$

where A is the contact area, C_0 is the abrasive concentration, and ϕ is the particle diameter. Since smaller particles have a larger surface area than larger particles, according to this model they will be able to remove more material than larger particles. A higher concentration of particles will also result in a larger contact area, thus leading to higher polish rates.

In contrast, according to the indentation-based mechanism, material removal during polishing occurs as a result of indentations created by the abrasive particles. The indent volume can be expressed as follows:

$$V \propto C_0^{-1/3} \cdot \phi^{4/3}$$

where V is the total indent volume. On the basis of this model, larger particles will have a larger indent volume, and consequently higher polishing rates. In addition, as the abrasive concentration (number of particles in contact with the wafer surface) increases, the force exerted

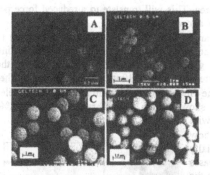

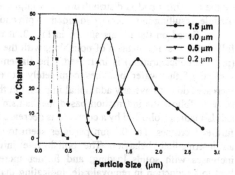

Fig. 1: A: TEM Micrograph of 0.2 μm silica particles, B: SEM Micrograph of 0.5 μm particles, C: SEM Micrograph of 1.0 μm particles and D: SEM Micrograph of 1.5 μm diameter particles

Fig. 2: Size distributions of particles measured by Dynamic Light Scattering

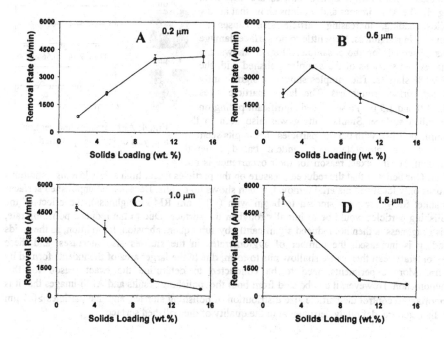

Fig. 3: Polishing rate as a function of solids loading for A: 0.2 μm B: 0.5 μm C: 1.0 μm and D: 1.5 μm diameter particles

by the polishing pad is distributed over a larger area. This will translate to a reduced force per particle, resulting in a decrease in indent depth and a lower polish rate.

Based on the results shown in Fig. 3, it can be seen that the 0.2 µm particles seem to follow the first mechanism of polishing, with the polishing rate increasing as the solids loading (particle concentration) is increased. The flattening out of the removal rate indicates that the surface of the wafer becomes completely covered at around 10wt.% loading, and further increases do not have any additional impact. On the other hand, the 1.0 µm and 1.5 µm particles seem to follow the indentation based mechanism, showing high removal rates for low (2wt.%) solids loading, followed by a continuous decrease with

further increases. The 0.5 µm particles seem to show an intermediate behavior. The polish rate initially increases with solids loading, and further increases lead to a reduction in removal rate, indicating that the removal mechanism in this case is probably a combination of the indentation and contact area models.

Atomic Force Microscopy (AFM) analysis was carried out on the above samples. The results for the samples polished with 2wt.% slurries are shown in Fig. 4. The AFM images and sections show that at low solids loading, increasing particle size causes an increase in roughness. Very little to no surface damage was observed for those samples. Fig. 5 shows the images and sections of the samples polished with the 15 wt.% slurries. The samples showed a significantly higher surface roughness. The larger particle sizes (0.5,1.0 and 1.5 µm) also caused significant pitting on the silica surface. Similar pitting was also seen in the sample polished with 0.5 µm particles. These pits seem to support our interpretation of indent based material removal, but the exact reason for their occurrence is not

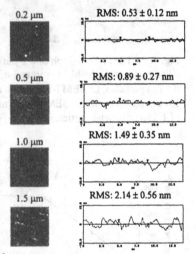

Fig. 4: AFM images and section analysis of samples polished with 2 wt.% slurries

clear. Our belief is that the reduced pressure on the particles under high solids loading conditions causes only localized material removal, which shows up as pits. Because the unpolished surfaces (blanket SiO_2) are very smooth to begin with (<0.3 nm RMS roughness), the effect of the polishing particles would be to initially roughen the surface. Due to the reduced polishing rate, this roughness is then not reduced significantly by subsequent abrasion. In addition, as the solids loading is increased, the number of agglomerates in the slurries also increases, and these agglomerates can then cause shallow pits to form, due to the larger area of the indents formed by them. More experiments need to be conducted to determine the exact reason for this phenomenon. However, it can be said from both the polishing results and AFM images that it is important to control the particle size distribution of polishing slurries, and that particles ≥0.5 µm in diameter are definitely detrimental to the quality of the polished surface.

CONCLUSIONS

The effect of particle size on polishing rate and surface roughness was studied for silica abrasives polishing SiO$_2$ thin films. The results indicated two removal mechanisms, an indentation-based mechanism for large particles and a surface area based mechanism for small (submicron) particles. A transition in the mechanism of polishing was also observed as the particle size was increased, with 0.5 µm as the size range where the transition took place. An increase in surface roughness was also observed with increase in particle size. The surface pitting observed for high solids concentration slurries of the large particles indicates that 0.5 µm is the size beyond which significant damage occurs on the oxide surface. These results, in conjunction with future work, can help develop a workable model for the oxide polishing process.

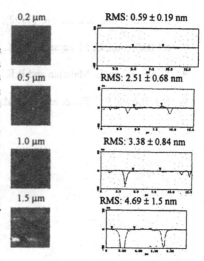

Fig. 5: AFM images and section analysis of samples polished with 15 wt.% slurries

ACKNOWLEDGEMENTS

The authors would like to acknowledge the financial support of the Engineering Research Center (ERC) for Particle Science and Technology. Thanks are also due to Seung-Mahn Lee for his help with the TEM.

REFERENCES

1. J.M. Steigerwald, S.P. Murarka and R.J. Gutmann, *Chemical Mechanical Planarization of Microelectronic Materials*, John Wiley and Sons, New York (1997)

2. L.M. Cook, J. Non-Cryst. Solids **120**, p. 152 (1990)

3. T. Izumitani, in *Treatise on Materials Science and Technology*, eds. M. Tomozawa and R. Doremus, Academic Press, New York (1979), p. 115

4. R. Jairath, M. Desai, M. Stell, R. Tolles, and D. Scherber-Brewer, in *Advanced Metallization for Devices and Circuits-Science, Technology and Manufacturability*, edited by S.P Murarka, A. Katz, K.N. Tu and K. Maex (Mater. Res. Soc. Proc. **337**, Pittsburgh, PA 1994), p. 121

5. Y. Xie and B. Bhushan, Wear **200**, p. 281 (1996)

6. S. Sivaram, M. H.M. Bath, E. Lee, R. Leggett, and R. Tolles, Proc. SRC Topical Research Conference on Chem-Mechanical Polishing for Planarization, SRC,

Research Triangle Park, NC (1992), proc. Vol. #P92008

7. M. Bielmann, U. Mahajan and R.K. Singh (this symposium)

8. M. Bielmann, U. Mahajan and R.K. Singh, J. Electrochem. Soc. (to be submitted)

9. N.J. Brown, P.C. Baker and R.T. Maney, Proc. SPIE **306**, 42 (1981)

A STUDY OF THE PLANARITY BY STI CMP EROSION MODELING

K.H. KIM, S.R. HAH, J.H. HAN, C.K. HONG, U.I. CHUNG and G.W.KANG
U-Tech. Team, Semiconductor R&D Center, Samsung Electronics Co. Ltd, San24 Nongseo-Ri Kiheung-Eup Yongin-City, 449-900, Korea, hyuncmp@samsung.co.kr

ABSTRACT

In this work, we propose a new equation that predicts the planarity as a function of active pattern density, initial step height, selectivity between gapfilled oxide and silicon nitride and over CMP amounts. In order to achieve highly planarized STI surface, uniform active density, reduced initial step height, minimization of over CMP amounts and high selective slurry were required. Our new equation was applied to the 0.18um graded CPU devices' STI CMP to enhance planarity and these parameters were evaluated quantitatively. It is concluded that the model suggested is useful in predicting CMP planarity

INTRODUCTION

Shallow Trench Isolation(STI) gives an improved isolation in the sub-micron devices, greater packing density and a superior planarity when compared to conventional isolation schemes. The high degree of planarity is essential to meet the Depth Of Focus(DOF) requirements with decreasing line width in the sub-quarter micron regime. Therefore Chemical Mechanical Polishing (CMP) is crucial in achieving highly planarized STI surface[1].

Unlike DRAM, the 0.18um grade Central Process Unit(CPU) has non-uniform in-chip active densities since it is comprises of capacitor, logic, high speed SRAM and Test Element Group(TEG) areas as shown in Fig.1 . Uneven pattern densities cause large SiN thickness variations within a chip after the STI CMP Process. These variations generate irregular step height differences between field and active regions and cause device failures due to gate oxide pitting and the gate poly silicon residues at the border.

In the present study, we propose a new equation to predict the planarity after STI CMP process by erosion modeling which incorporates active pattern density, initial step height, selectivity between oxide and silicon nitride and over CMP amounts.

MODELING

For simplicity and to understand the STI mechanism, we introduce the following assumptions. First, the planarization length is zero, that is, there is no interaction between removal rates of patterned and blanket areas, and second, there is no dishing or recess at field oxide between active silicon nitrides in feature size level. On the basis of the above assumptions, STI CMP procedures can be divided into four steps as shown in Fig.2. The first step is defined as the period in which initial step heights of patterned area are perfectly eliminated. At this stage then erosion is generated due to the difference of removal rate between patterned and blanket area as shown in Fig.2(b). The second step is defined as the period in which the fully planarized oxide surface of patterned area is polished to expose the silicon nitride top surface.

33

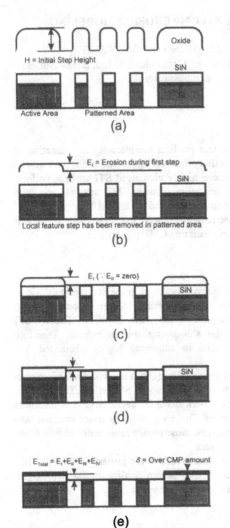

(a)

H = Initial Step Height

Oxide

SiN

Active Area   Patterned Area

(b)

E_I = Erosion during first step

SiN

Local feature step has been removed in patterned area

(c)

E_I ($\because E_{II}$ = zero)

SiN

(d)

SiN

(e)

$E_{Total} = E_I + E_{II} + E_{III} + E_{IV}$ δ = Over CMP amount

Figure 2. The schematic diagram for STI CMP process.

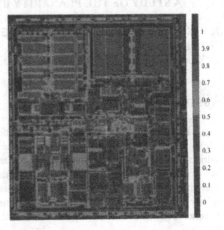

Figure 1. Active pattern density Map of 0.18um grade CPU device.

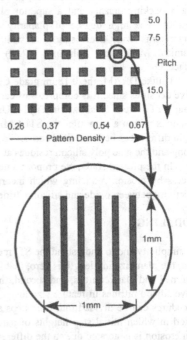

Pitch

5.0
7.5
15.0

0.26   0.37   0.54   0.67

Pattern Density

1mm

1mm

Figure 3. Die layout of varying pattern densities and pitch sizes to test removal rate ratio of the oxide/nitride mixed area.

In this period, there is no newly generated erosion because of the same removal rates in both areas as shown in Fig.2(c). During the third step, oxide layer that remains after the second step is polished in blanket area, whereas mixture of oxide and nitride is polished in the patterned area. Because the removal rate of the oxide/nitride mixed area is smaller than that of oxide, the erosion that is generated in first step decreases with polishing time during the third step. The final step is defined as the period in which erosion is generated due to the difference of removal rate between the mixed area and the nitride area as shown in Fig.2(d). In order to quantify and elucidate the STI CMP as explained above, some simple relations which can predict the post STI CMP erosion were derived.

For simplicity, we define the oxide removal rate ratio, N_{OX}, as follows ;

$$N_{OX} = \frac{V_{OX}}{V_{OX}^0} = \frac{V_{OX}^0/\rho_E}{V_{OX}^0} = \frac{1}{\rho_E}$$

(1)

where V_{OX} is the oxide removal rate of patterned area, V_{OX}^0 is the removal rate of blanket oxide layer and ρ_E is the effective oxide density[2,3]. The effective oxide density is defined as the ratio of deposited oxide volume to unit volume, and it is strongly dependent on the kinds of oxide film, deposition methods and oxide etch back schemes.

In order to obtain an expression for the erosion amount in first step, E_I, at various densities, the following relation[4] is used ;

$$E_I = t_I V_{OX} - t_I V_{OX}^0 = t_I V_{OX}^0 (N_{OX} - 1)$$

(2)

where t_I is the time of the first step. t_I is related to the initial step height, H, and the oxide removal rate of patterned area, V_{OX}, by the following relationship ;

$$t_I = H / V_{OX}$$

(3)

From eqns.(1)-(3), the following relation is obtained for the erosion needed to eliminate the local oxide step height in the patterned area ;

$$E_I = t_I V_{OX}^0 (N_{OX} - 1) = \frac{H}{N_{OX} V_{OX}^0} V_{OX}^0 \left(\frac{1}{\rho_E} - 1\right) = H(1 - \rho_E)$$

(4)

The values for the erosion is zero and H(initial step height) when effective oxide density of the patterned area is 1 and zero, respectively. The erosion increases with increasing initial step height and decreasing effective oxide pattern density during the first step. The effective oxide density of second step becomes one because the local oxide step height in patterned area was fully eliminated during the first step ;

$$E_{II} = t_{II} V_{OX} - t_{II} V_{OX}^0 = t_{II} V_{OX}^0 - t_{II} V_{OX}^0 = 0$$

(5)

The erosion amount in the third step may be expressed by the following equation ;

$$E_{III} = t_{III} V_{MIX} - t_{III} V_{OX}^0 = t_{III} V_{SiN}^0 N_{MIX} - t_{III} V_{OX}^0 \qquad (6)$$

where, V_{MIX} is the removal rate of nitride/oxide mixed area and V_{SiN}^0 the removal rate of blanket silicon nitride layer and t_{III} is the time of the third step. Since silicon nitride and gap filled oxide are polished simultaneously in the third step of STI CMP as shown in Fig.2, it is very important to evaluate the removal rate of the mixed area. The removal rate of nitride and oxide mixed area can be expressed by the removal rates of blanket oxide and nitride through the parallel rule of mixtures[5] ;

$$V_{MIX} = \frac{V_{OX}^0 \, V_{SiN}^0}{\rho \, V_{OX}^0 + (1-\rho) V_{SiN}^0} \qquad (7)$$

Therefore we can define the nitride and oxide mixed area removal rate ratio, N_{SiN}, as follows ;

$$N_{SiN} = \frac{V_{MIX}}{V_{SiN}^0} = \frac{V_{OX}^0/V_{SiN}^0}{\rho\left(V_{OX}^0/V_{SiN}^0\right)+(1-\rho)} = \frac{S}{\rho(S-1)+1} \qquad (8)$$

From the relationship between E_I, V_{OX}^0 and t_{III}, the following equation can be obtained because time of the third step is equal to the time to eliminate the remnant oxide on the blanket nitride from the first step ;

$$E_I = t_{III} V_{OX}^0 \qquad (9)$$

From eqns.(6)-(9), the following relation is obtained for the third step of STI CMP as a function of initial step height, oxide effective density, active pattern density, ρ, and selectivity of slurry, S ;

$$E_{III} = \frac{V_{SiN}^0 E_I S}{V_{OX}^0 \rho(S-1)+1} - E_I = \frac{E_I}{\rho(S-1)+1} - E_I = \frac{H(1-\rho_E)}{\rho(S-1)+1} - H(1-\rho_E) \qquad (10)$$

The erosion amount in the fourth step may be expressed by the following equation using eqns.(7) and (8) ;

$$E_{IV} = t_{IV} V_{SiN} - t_{IV} V_{SiN}^0 = t_{IV} V_{SiN}^0 N_{SiN} - t_{IV} V_{SiN}^0 = t_{IV} V_{SiN}^0 \frac{(1-\rho)(S-1)}{\rho(S-1)+1} \qquad (11)$$

The definition of over-CMP amounts, δ, is as follows ;

36

$$\delta = t_{IV} V_{SiN}^0 \tag{12}$$

Using eqns.(11) and (12),   the following relationship is obtained ;

$$E_{IV} = \delta \frac{(1-\rho)(S-1)}{\rho(S-1)+1} \tag{13}$$

Total erosion of STI CMP is the sum of erosions generated during each steps, and is expressed as following equation ;

$$E_{Total} = \sum_{i=I}^{IV} E_i = \frac{H(1-\rho_E) + \delta(1-\rho)(S-1)}{\rho(S-1)+1} \tag{14}$$

The final expression to predict the erosion during STI CMP becomes the function of initial step height, effective oxide density, over CMP amount, active pattern density and selectivity between oxide and nitride.

RESULTS

In order to compare the mechanical behavior of the mixed region with proposed and experimental data, we investigated the effect of active pattern density and pitch size on the removal rate of mixed area. Fig.3 shows the layout pattern used to estimate the removal rate of mixed area. The mask consists of 48 1mm×1mm density structures in a die. The local pattern density, defined as the ratio of line width to pitch, is gradually increased from 25% to 65% along with horizontal axis and the pitch is gradually increased from 2.5um to 15um along the vertical axis. Patterned SiO$_2$/SiN/Si layer with respective trench depths of (8K/2K/4K) was polished and step height with polishing time was measured using the P-2h profilemeter. In this experiment all polishing process was carried out on a 6DS-SP polisher. IC1000/Suba IV k-grooved pads and SS-25 slurry were used on the primary platen.

Fig.4 shows the calculated and experimental removal rate ratio of the mixed area at various pattern densities and pitch sizes. The calculated data increases with decreasing active pattern density and it is in good agreement with the experimental data. As can be seen in the diagram blanket oxide to nitride selectivity is 4.3 and decreases as active density increases. The selectivity change is greater in the low active density areas than in the high density areas.

Fig.5 shows the effect of initial step height on erosion at various active densities. Erosion increases linearly as step height is increased from 2000Å to 6000Å. At 40% active density, reducing the initial step height from 6000Å to 2000Å reduces the erosion amount by 1000Å. But at 60% active density, reducing the initial step height from 6000Å to 2000Å reduces the erosion amount by 500Å. This implies that erosion and planarity can be improved by reducing the initial step height through a pre-cmp photo + oxide etchback step. An improvement in planarity was achieved by applying a pre-cmp photo + oxide etchback step on a CPU device.

Fig.6 shows the effect of over CMP on the erosion at various active densities. Erosion due to differences in pattern density increases with SiN polishing amount. Therefore reducing

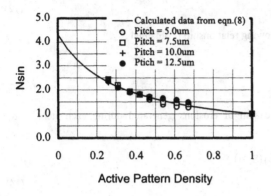

Figure 4. Measured removal rate ratio of oxide/nitride mixed area at various densities and pitches along with values calculated from equation (8). (S = 4.3)

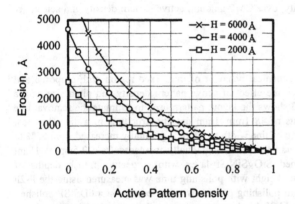

Figure 5. The effect of initial step height on the erosion at various active pattern densities. (δ=200Å, S=4.3, ρ_E=ρ)

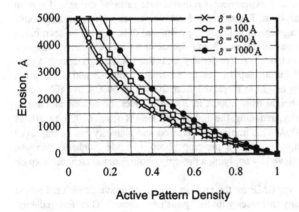

Figure 6. The effect of over CMP amounts on the erosion at various active pattern densities. (H = 6000Å, S = 4.3, $\rho_E = \rho$)

the SiN overpolishing amount is recommend for erosion reduction and especially stresses the importance of dummy features to reduce in-chip SiN variation which can overcome the erosion increase at low active densities.

Fig.7 shows the effect of slurry selectivity between oxide and nitride on the erosion at various densities. Slurry plays an important role in STI CMP. As can be seen in the diagram slurry selectivity has a big effect on the amount of erosion. At high slurry selectivities, erosion can be kept at a minimal level for high active densities. But as can be seen in Figure 6, erosion increases rapidly at low active densities. Therefore to minimize erosion, a slurry with optimum selectivity for a given active density and density distribution needs to be used. In-chip planarity after STI CMP can be simply expressed by the following equation ;

$$P = \Delta E = E^{MAX} - E^{MIN} = \left| \frac{H(1-\rho_E) + \delta(1-\rho)(S-1)}{\rho(S-1)+1} \right|_{MIN}^{MAX}$$

$$= \frac{H\left(1-\rho_E^{MIN}\right) + \delta\left(1-\rho^{MIN}\right)(S-1)}{\rho^{MIN}(S-1)+1} - \frac{H\left(1-\rho_E^{MAX}\right) + \delta\left(1-\rho^{MAX}\right)(S-1)}{\rho^{MAX}(S-1)+1}$$

The effect of slurry selectivity on planarity for various active density was investigated by the following method. If we assume the active density varies by 30%, the dependence of planarity on slurry selectivity is shown in Fig.8. For selectivity value 1, planarity value 1800Å is not affected by the active density. On the other hand when the slurry selectivty increases above 1, planarity is affected by the active density. For active densities of 45±15% or greater, post-CMP planarity decrease with increasing slurry selectivity. But chips with active densities of 35±15% and 25±15% show planarity increases with increasing selectivity until a certain value where it decreases. For active density of 25±15% better planarity was achieved with selectivity of 1 than with selectivity of 32. This kind of behavior is also affected by the over-CMP amount and Fig.9 shows the effect of slurry selectivity on planarity at 500Å overpolish. In Fig. 8, 45±15% group, that showed planarity decrease with increasing selectivity, has worst planarity at selectivity 4. And also for 25±15% active density group, planarity increased for selectivities 1~32.

Fig.10 shows the CMP surface height for a CPU device with polishing time for various conditions. Planarity is defined as the height difference between the different regions. Fig. 10(a) is the results for active dummy insertion and selectivity of 4.3 for STI CMP. Chip planarity decreases rapidly in the early stages of CMP and stays at 300Å with the exposure of SiN surfaces but increases with polishing due to the active density differences. Under identical conditions using a slurry with selectivity of 45 reduces the planarity to 100Å at SiN exposure(Fig.10(b)). But for the case of no active dummy as in Fig.10(c),(d), CMP erosion increases rapidly in the low active density areas and increases the planarity. This increase in erosion and planarity is amplified for high selectivity case.

CONCLUSIONS

A quantified model that explains planarity characteristics is established and this in turn make it possible to identify and quantify various planarity factors. The proposed model allows one to represent erosion as functions of initial step height, slurry selectivity, over-CMP

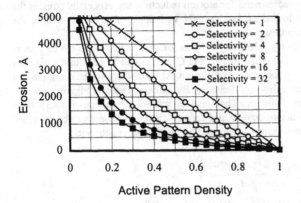

Figure 7. The effect of slurry selectivity between oxide and silicon nitride on the erosion at various active pattern densities. (H = 6000Å, δ = 200Å, $\rho_E = \rho$)

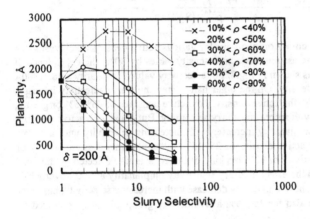

Figure 8. The effect of slurry selectivity between oxide and silicon nitride on the planarity at over CMP amount = 200 Å for various active pattern density distributions. (H = 6000Å, $\rho_E = \rho$)

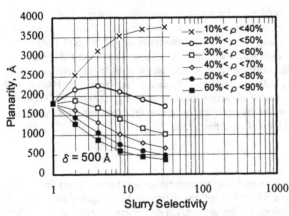

Figure 9. The effect of slurry selectivity between oxide and silicon nitride on the planarity at over CMP amount = 500 Å for various active pattern density distributions. (H = 6000Å, $\rho_E = \rho$)

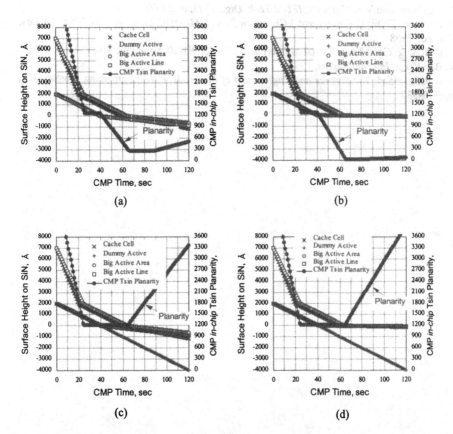

Figure.10 Chip planarity variation at various active dummy conditions and slurry selectivities of (a) with dummy and selectivity of 4.3 , (b) with dummy and selectivity of 45, (c) without dummy and selectivity of 4.3 and (d) without dummy and selectivity of 45.

amount and effective density. The result based on our model suggested requires decreased step height, uniform active density and minimal amount of over-CMP to reduce CMP erosion.

REFERENCES

1. P.C.Fazan and V.K.Mathews, *IEDM Tech. Dig.*, 1993, pp.57-60
2. D.Ouma, C.Oji, D.Boning and J.Chung, *in Proc. CMP-MIC Conf.*, pp.20-27, Santa Clara, CA, Feb 1998
3. D.Boning, *CMP Tech. For ULSI Interconnection*, pp.F1-20, San Francisco, CA, Jul. 1998
4. B.U.Yoon, *Internal Report*, 1998
5. D.R.Askeland, *The Science and Engineering of Materials*, PWS Engineering, Boston, MA, pp.373-387

Part II

Pads and Related Issues

Modelling the influence of pad bending on the planarization performance during CMP

Joost Grillaert, M. Meuris, E. Vrancken, N. Heylen, K. Devriendt, W. Fyen, M. Heyns.
Imec, Belgium.

Abstract

One of the major problems with oxide-CMP is the oxide thickness variation after CMP within one die, the socalled WithIn Die Non-Uniformity (WIDNU). The variations in pattern density of the design layout causes different local removal rates across the die resulting in the WIDNU. In this paper we shown that this is influenced by the pad stack. Depending on the thickness of the top pad the WIDNU can be reduced from about 470 nm until almost zero. This will be related to the bending of the top pad. The modelling will focuss on the two extreme cases of perfect pad bending and no pad bending.

Introduction

Chemical mechanical polishing planarizes efficiently structures up to 1 mm. On one hand this short range planarity results in small step heights approaching zero within one structure on a wafer polished for a long time. On the other hand global planarity requires a flat surface within one die. This long range planarity up to the die size (2-3 cm) is often referred to as the WithIn Die Non Uniformity (WIDNU) for the oxide, i.e. the range of oxide thickness within one die. Typical values range between 200 to 400 nm. This is to large for the subsequent lithography steps.

In ref. 1-6 it is concluded that the pattern density is the dominant layout factor contributing to the WIDNU. A model is presented for the dependence of the local removal rate on the pattern density. Based on this model the oxide thickness as function of the polishing time is calculated and the WIDNU is deduced. Generally a stacked pad is used during CMP: a hard top pad is glued on a soft bottom pad. Two extreme cases are considered: the top pad bends perfectly or there is no bending at all. To our knowledge up to now only perfect pad bending was assumed in the literature (see e.g. ref. 3 and 4).

In the literature several models are presented that try to predict the oxide thickness after CMP at each point starting from the layout of the design (ref. 3 and 6). This is of course the ultimate target, but in order to obtain a reasonable agreement between the model and the experiment several fit parameters without physical interpretation are introduced. The purpose of the model presented in this paper is not to calculated the oxide thickness after CMP at each location of the die, but to give an understanding of the importance of the pad bending and to identify the key parameters. With the obtained insight further optimization of the CMP-process is possible.

Experimental setup

The test wafers had the following process flow. After the deposition of 200 nm of oxide the metal stack with a total thickness of 2200 nm was sputtered. The wafers

45

were patterned with a special test mask with structures of various pattern density (0 %, 25 %, 50 % and 75 %). The pitch of these structures was 400 or 800 μm allowing optical measurements on a KLA-Tencor FT-700. After metal etching a thick oxide of 4000 nm was deposited. The wafers were polished on an Ipec Avantgaard 472 and cleaned on an Ontrak scrubber. The thickness of the top pad was varied between 32, 50 and 100 mils. For one condition a 50 mils thick top pad was put immediately on the platen.

Description of the model
Pad bending
At the beginning of the CMP-process the average pressure on an unit area is uniform and equal to the nominal down force. This results in an uniform average deformation of both the top and the bottom pad. The dependence of the polishing on the pattern density causes variations of the local removal rate across the die. As a consequence level differences are introduced: the level of the up features of the dense structures will be higher than the one of the isolated structures. Since the silicon surface and the polisher platen remain parallel during the entire process these level variations will cause deformations of both the top and the bottom pad. Which one is deformed most is determined by the bending of the top pad.

If the top pad is regarded as a beam than the maximum bending (u) depends on the pressure (P), the distance between the two supporting points (L), the Young modulus (E) and the thickness of the beam (b), i.e.,

$$u = \frac{60}{384} \frac{PL^4}{Eb^3} \tag{1}$$

Figure 1 illustrates the influence of the pad thickness and of the distance between the support points on the pad bending using eq. 1. The distance between the support points was varied from 0 till 5 mm. Since a chip size is typical 1-4 cm^2 these distances are common on designs. Based on the bending of the pad two extreme cases are considered: perfect pad bending and no pad bending. Perfect pad bending assumes that the WIDNU is completely compensated by the deformations of the bottom pad. In the no pad bending case the level differences within one die cause only additional deformations of the top pad.

Extreme case 1: perfect pad bending
Suppose a thin hard pad is on top of a thick soft pad. In this case it is concluded from figure 1 the top pad bends easily. As a consequence the bottom pad will be deformed non uniformly. Typical values for the bottom pad are a Young modulus (E) of 20 MPa and a thickness (l) of 1 mm. The maximal change in thickness for the bottom pad can be deduced from the results for the within die non-uniformity. A typical value is 300 nm. In the perfect pad bending assumption this level difference is entirely translated into thickness variations of the bottom pad (Δl). Hooke's law allows to calculate the additional pressure resulting from these deformations, i.e.

$$P = E \frac{\Delta l}{l} \tag{2}$$

This results in a pressure difference of 0.87 PSI. Compared to a nominal down force of 7 PSI this can be neglected in a first order approximation. This leads to the

46

fundamental assumption of the perfect pad bending case: the average total local pressure should be the same for all unit square areas (e.g. 1 mm²), equal to the nominal tool setting (e.g. 7 PSI) and independent of polishing time or level differences.

In reference 1-7 it was shown that the local removal for the up features was equal to the removal rate on a flat wafer (r) divided by the pattern density (a). If the pad surface could be modeled as incompressible this remains valid till the structure is completely planarized at the planarization time t_{pa}. Afterwards the level is reduced at the same rate as the polishing rate on a flat wafer. Since the average pressure is always the same for an unit area in the case of perfect pad bending this calculation can be applied for all structures independent of the surrounding areas. This allows to calculate the WIDNU as the level difference between the area with the highest pattern density (b) and the lowest pattern density (e.g. a), i.e.

$$WIDNU = h_0(b - a) \tag{3}$$

This means the WIDNU is not zero in the case the pad bends perfectly like a thin top pad on a soft bottom pad. In reference 7 the WIDNU was also calculated in the case of a compressible pad surface, but the equation for the WIDNU remained the same as in the case of an incompressible pad surface. That is why for the no pad bending case the pad surface will be assumed to be incompressible.

Extreme case 2: No top pad bending

A practical example of this extreme case is when the top pad is put immediately on the platen of the polisher. According to figure 1 the bending is also limited if a thick top pad is put on a soft bottom pad and if the distance L is smaller than a few millimeter. In this case the level differences between different structures after a short polishing time introduce an additional deformation, not at the surface, but in the bulk of the pad. An example can show that even small level differences cause a significant increase or decrease in the pressure on a certain area. Suppose the level difference is 100 nm, the Young Modulus is 366 MPa and the thickness of the top pad is 1 mm. For these typical values Hooke's law gives an additional pressure of 5.3 PSI to the local pressure of the surface pad deformations. This value is almost comparable to the standard setting of 7 PSI down force.

The bulk deformation at a certain point will be influenced by the topography of the surrounding area. For the modeling only the interaction between an area with pattern density a and an area with pattern density b is considered. ΔH denotes the level difference of their up features. The no pad bending assumption implies this is also the difference of the bulk deformation between area a and b. Compared to the start of the polishing process the bulk deformations are changed. At the location with the lowest pattern density (a) it is assumed the bulk deformation is decreased by half of ΔH and at location with the highest pattern density (b) it is increased by half of ΔH. Substituting Hooke's law in the Preston equation allows to correlate ΔH with the change in local removal rate.

$$\Delta r_l = KVP = KVE \frac{\Delta H}{2l} = \frac{\Delta H}{2\tau} \tag{4}$$

The local removal rates are now given by $r_a = \frac{r}{a} - \frac{\Delta H}{2\tau}$ and $r_b = \frac{r}{b} + \frac{\Delta H}{2\tau}$ (5) and (6)

The local removal rates are the derivatives towards time of the different levels and the ΔH is the difference between these levels. This results in a set of coupled differential equations.

$$\frac{dd_{au}}{dt} = -\frac{1}{2\tau}d_{au} + \frac{1}{2\tau}d_{bu} - \frac{r}{a} \qquad \frac{dd_{bu}}{dt} = +\frac{1}{2\tau}d_{au} - \frac{1}{2\tau}d_{bu} - \frac{r}{b} \qquad \text{(7) and (8)}$$

Solving this set of equations gives the levels of the different structures as a function of time. This set can be solved analytically, but in this paper only a quantitative discussion will be presented.

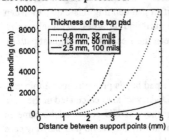
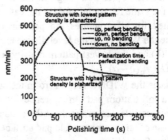

Figure 1 (left): The bending of the top pad as a function of the distance between the supporting points and for various pad thickness.

Figure 2 (right): The local removal rate simulated for a structure with a pattern density of 75 % in the case of perfect bending and no bending

At a certain polishing time a structure will be planarized. At that moment its pattern density becomes 1 in the differential equation. Guaranteeing the continuity of the oxide levels results in the initial conditions for these sets of differential equations.

Figure 2 shows the simulated local removal rate for an area with a pattern density of 75 % for the different phases. They were obtained by a derivation of the different oxide levels. As comparison the removal rates in the case of perfect pad bending is included. The removal rate on a flat wafer was assumed to be 220 nm/min, the time constant were assumed to be 50 s.

Initially the local removal rate increases until the planarization time for structure a. is Because the level of the dense area is higher than the other, the bulk deformation is larger resulting in an increased pressure. At the planarization time for structure a the local removal for structure a is reduced. As a result the level differences and the bulk deformations are reduced. This results in a drop of the removal rate. When structure b is completely planarized, the removal rate is discontinue like in the case of perfect pad bending. Afterwards the local removal rate for structure b drops gradually till its final value.

The local removal rate for the area with the lowest pattern density is according to the simulation lower in the no pad bending assumption than in the perfect pad bending assumption. This is related to the lower bulk deformation for this structures.

According to the simulations the WIDNU in the no pad bending assumption approaches zero after a long polishing time. This should be opposed to equation 3 in the perfect pad bending case. This is due to both the higher removal rate for the dense structure and the lower removal in the isolated structure compared to the perfect pad bending assumption.

Results

The area with the largest pattern density has always the highest level. Figure 3 shows the results for the up features for the area with a pattern density of 75 %.

The first measurement point for the local removal rate is about 440 nm/min for the pad without a bottom foam. It starts to increase to a maximum value of 550 nm/min. After this maximum value it drops to 370 nm/min. When the removal rate starts to drop the down features of the area with a high pattern density do not yet support the applied load because of the high step height (2300 nm). This is evidenced by the fact no oxide is removed in the down areas of this high density structure. The drop in the removal rate can only be explained by the interaction with the surrounding areas. Their pattern density is lower, the pad deformations in these areas start to contact the down features and they start to support the applied load. This reduces the local removal rate of the surrounding areas, the level differences between the different areas decrease and the bulk deformations become smaller in the area with the highest pattern density. This results in the observed decrease in the removal rate.

It is obvious that the no pad bending model is applicable if the top pad is put immediately on the platen. The results show however that a 100 mils thick top on a soft bottom pad has a rather similar behavior. This shows that a thick top pad can suppress the pad bending. Note the similarity between the simulations of figure 2 for the no pad bending case and the experimental results of figure 3.

For the thinnest top pad the removal rate starts at a comparable level to that for the thick pad, but the evolution as a function of the polishing time is different. It remains relatively constant for the rest of the polishing time. Only a small increase towards the end of the polishing time is observed. These observations indicate that the perfect pad bending model is more applicable for this case. Also in this case the simulations for perfect pad bending and the experimental results are in good agreement.

For the top pad with the "in-between thickness" (50 mils) the removal rate increases initially slightly, reaches its maximum and then it drops as well. Its maximum value is smaller than for the thick top pad. Based on figure 3 it is concluded this pad bends more than the thick top pad. The bulk deformations are smaller and the observed increase in local removal rate is lower.

The local removal rate in the area with a pattern density of 0 % is shown on figure 4. For the pad without the foam the removal rate is initially low. As polishing is proceeding the removal rate increases gradually to the value on a flat wafer. Initially the up features of the surrounding areas of this large open field support the pad. Since no foam is present bending is suppressed. The large area without any structure is not contacted because of the high initial level difference. As this level difference becomes smaller the large area without up features starts to support the applied load and oxide is removed, resulting in the gradual increase of the removal rate.

On the other hand the oxide removal in the area starts immediately when a thin top pad is used (figure 4). The local removal rate is close to one on a flat wafer. It remains high, but a small, not expected decrease for the longer polishing times is observed. The up features of the surrounding areas support still the pad, but it is clear that the pad bends in the open area and the latter supports the pad as well: the area is polished like it is a flat wafer.

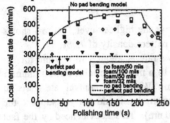 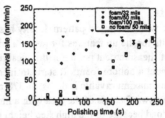

Figure 3 (left): The experimental local removal rate for a structure with a pattern density of 75 % as function of the polishing time for different pad stacks.

Figure 4 (right): The experimental removal rate for a large area without up features as a function of the polishing time for different pad stacks.

The different evolution of the local removal rate as function of the polishing time results in different WIDNU-results for the various pad stacks. The WIDNU is the highest for the thin top pad on the bottom pad (474 nm) and reduces for the thick top pad on the bottom pad until 164 nm . The WIDNU was the smallest if no bottom pad is used (17 nm).

Conclusions
The pad stack determines the bending of the top pad. Two extreme case are considered: perfect pad bending and no pad bending. A typical example for perfect pad bending is when a thin top pad is put on a soft bottom pad. In this case the top pad bends easily and it was concluded that the average pressure on an unit area is equal to the nominal pressure. This results in a large oxide thickness variation after CMP. In the case of a thick top pad on a soft top pad the top pad does not bend. This introduces bulk deformations in the top pad. In this case the oxide thickness variation after CMP is limited. The experiments confirm that a real stacked pad is always in between these two extreme cases.

References
1) R.R. Divechaet al. , CMPMIC 1997, p. 29.
2) J. Grillaert et al., Advanced Metallization and Interconnect Systems for ULSI Applications in 1996, MRS, p. 525 (1997)
3) B.E. Stine et al., CMP-MIC 1997, p. 266.
4) B.E. Stine et al., IEEE Transactions on Semiconductor Manufacturing, vol. 11, no.1, 129 (1998).
5) R.R. Divecha et al., J. Electrochem. Soc. , 145, 1052 (1998)
6) J. Warnock, J. Electrochem. Soc. , 138, 2398 (1991)
7) J. Grillaert et al., CMPMIC-97, p.

STRUCTURED ABRASIVE CMP:
LENGTH SCALES, SUBPADS, AND PLANARIZATION

D. P. GOETZ
3M Advanced Materials Technology Center
Building 201-1W-28, 3M Center, St. Paul, MN 55144-1000, dpgoetz@mmm.com

ABSTRACT

Chemical-Mechanical Planarization with structured abrasive uses a subpad to manage the pressure variations due to loading over a range of length scales. The effect of subpad construction on pressure responses related to those scales is illustrated.

A minimum length scale for the effect of the subpad is established via contact mechanics. Differences between one- and two-layer subpads are shown. Uniform compression, point loading, and edge exclusion are considered briefly. A model of the subpad as a plate on an elastic foundation is applied to the problem of die doming. The roles of process pressure, die size, and subpad construction are illustrated. Planarization at the intra-die, die, and wafer scales are related to the subpad construction.

INTRODUCTION

Chemical-Mechanical Planarization involves often-conflicting requirements at various length scales--e.g. uniform removal at the wafer scale, but non-uniform removal of high areas to achieve planarization at the feature scale. In conjunction with machine process controls, the management of pressure by the consumables is one key to balancing these requirements.

3M is developing abrasive systems for CMP in which the abrasive particles are incorporated into resin-based structures of precise dimensions on the surface of polymeric webs [1-3]. In order to manage pressure, the structured abrasive web can be laminated to a multilayer subpad. A typical construction is shown in Figure 1. The subpad is made up of the layers between the machine platen and the structured abrasive. Because the thin fixed abrasive layer constituting the process surface is independent of the subpad, there is great flexibility in subpad design. The primary mechanical elements in the pad can be changed while leaving the abrasive layer—and thus the details of the material removal process—unchanged, apart from the pressure distribution.

The objective of this study is to identify the controlling features of structured abrasive system performance relative to the length scales of concern. The focus is on the role of the

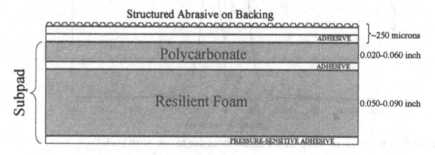

Figure 1. Schematic of typical construction of pad for slurry-free CMP showing subpad

subpad. The aim is to use ideas from applied mechanics to illustrate the relative effects of the subpad construction for the intra-die, die, and wafer scales. Such a physics-based understanding of the interaction of subpad constructions and length scales can inform the choice of subpad construction for experiments to optimize the CMP process.

THEORY

Load Transfer into the Subpad

There is a minimum length scale for loading of the subpad. Consider a point asperity contacting the surface of the abrasive layer. As the related stress transfers through the abrasive layer, it spreads out, so that the subpad "sees" the point load on the surface of the abrasive as a pressure distribution. Consider two point loads on an elastic half space as shown in Figure 2. The resulting normal pressure profiles along two sub-surface planes has been determined using the published results for stresses due to point loading [4]. If the layers above the subpad are 250 microns thick (see Figure 1), the normal pressure distribution on the subpad due to each point load covers a circle of roughly 1 mm in diameter. Along a plane 250 microns below the surface, the resulting normal pressure distributions from the two loads overlap, but are still resolvable when the spacing of the loads is 500 microns. As the spacing is decreased, the two distributions will merge. As this takes place, the subpad is no longer able to modify the pressure at one point preferentially. At the sub-millimeter scale, one would expect the subpad to have a rapidly diminishing effect on the pressure distribution at the process surface. At 1000 microns below the surface, the normal pressure distribution is seen to be a single peak. The thickness of the abrasive layer determines the length scale at which the subpad can resolve features at the abrasive surface and modify the pressure distribution. In experiments, one can expect the subpad to have influence on planarization primarily at the multi-millimeter scale because of the way loads are introduced into the subpad.

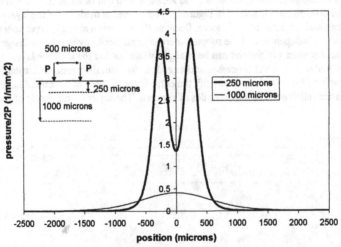

Figure 2. Sub-surface normal pressure distributions due to two point loads

Single-layer vs. Two-layer Subpads

Planarization requires preferential material removal at areas that are high relative to the surrounding surface. Because a pad is not completely rigid, the pressure at a particular point will depend on the topography of the wafer surface in a neighborhood around the point. Single-layer and multilayer pads differ fundamentally in the mechanics determining the size of the neighborhood. To illustrate the difference, pad surface deformations from a localized surface displacement are shown in Figure 3 for one- and two-layer subpads. An axisymmetric finite element analysis was used. The single layer subpad is a 0.02 inch thick sheet of polycarbonate. The radius of the non-zero surface displacement is about 500 microns. If another point displacement were located 1000 microns away, the associated pressure at the first point displacement would not be modified by its presence. The "interaction distance" is determined by contact mechanics. The two-layer subpad is a sheet of 0.06 inch thick polycarbonate on a 0.09 inch thick layer of polymeric foam. (Hereafter subpad constructions will be designated by, for example, PC60/90.) The surface deflection profile shows two regions. Close to the point of load application, the shape of the surface echoes that of the single layer subpad. This contact deflection is superimposed on a much longer-scale deflection profile. The longer-scale deflection is due to the bending of the top layer. The size of the neighborhood is on the order of 20 mm in radius. Plate bending mechanics rather than contact mechanics dominate.

The deflections are normalized in Figure 3. The stiffness of the single layer subpad is much higher than the two-layer subpad. For an imposed displacement, the local pressure will be much higher for the single layer subpad. In general, one would expect a single layer subpad to planarize by producing higher local pressures that are much more localized whereas a two-layer subpad planarizes by referring the pressure at a point to the topography in a much larger neighborhood.

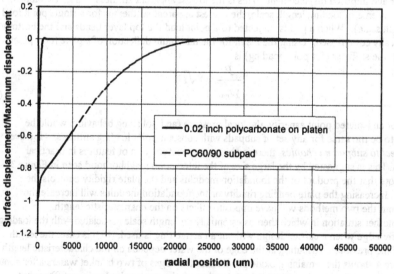

Figure 3. Surface displacements for point loaning of a one- and two-layer subpad at the origin

Plate on an Elastic Foundation Model

Features of the two-layer subpad response can be captured by considering the subpad as a plate on an elastic foundation. The governing equation for the deflection of a plate of bending stiffness D on a foundation of compression stiffness k is [5]

$$\lambda^4 \left(\frac{\partial^4}{\partial x^4} + 2\frac{\partial^4}{\partial x^2 \partial y^2} + \frac{\partial^4}{\partial y^4} \right) w(x,y) + w(x,y) = \frac{q(x,y)}{k}$$

(1)

where x and y are spatial coordinates, w is the plate deflection, q is the normal pressure on the plate, and λ is a characteristic length (units of length) given by

$$\lambda = \sqrt[4]{D/k}$$

(2)

The equation can be written in simpler form without introducing the characteristic length. The motivation for introducing it is that it appears as a constant in solutions to the governing equation. It is a parameter that combines the material properties and thicknesses of the layers to give a fundamental property of a subpad that sets the length scale of pad responses to pressure or displacement inputs.

The apparent stiffness of a pad depends on the length scale of loading. If a subpad is uniformly compressed (infinite length scale of loading), the top layer does not bend. By inspection, equation 1 reveals that the pressure depends only on the foundation modulus of the foam layer and imposed displacement. This is the lowest apparent stiffness of a given subpad. At the other extreme of loading length scale, the highest apparent stiffness is for a single point load (zero length scale). When a pad is subjected to a point load the top layer bends and the bottom layer is locally compressed. Using the result for the maximum deflection of a plate for a point load P [6], the stiffness (for point loading) is

$$\frac{P}{w_{max}} = 8\sqrt{Dk}$$

(3)

For an isolated point asperity, the local pressure (and polishing behavior) would be expected to be the same for any set of subpads with constant point loading stiffness. However *with respect to subpad mechanics*, there is always some interaction of features in practical situations. This is suggested by the large size of the interaction neighborhood seen in Figure 3.

Note that the product of the foundation modulus and the plate rigidity controls the stiffness. Increasing the plate bending rigidity or the foundation modulus will increase the stiffness, but the two methods will have opposite effects on the characteristic length.

Another situation in which there is essentially no length scale associated with the loading is at the wafer edge when a carrier design without an active ring is used. In this case, the length scaling of the pressure distribution at the edge of the wafer depends on the characteristic length [7]. Figure 4 shows the remaining oxide profile near the edges of two blanket wafers after about 5000 Angstroms of oxide were removed. Subpads with the same point loading stiffness but

54

different characteristic lengths were used. The oscillation in the profile reflects the oscillating pressure profile associated with the response of a plate on an elastic foundation.

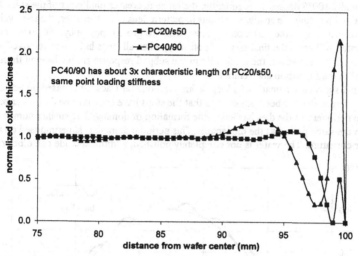

Figure 4. Remaining oxide thickness near the wafer edge for two subpads.

Die Doming

Die doming is sometimes encountered in practice. Because pattern density is often higher near the center of the die than at the edge, the remaining oxide thickness after removal of steps is thicker near the center, forming a "hill" or "dome". An important feature of die doming is that it is periodic, with a wavelength equal to the die size. As used here, "doming" will be used to refer to any periodic, symmetric oxide thickness variation resulting from an underlying pattern density variation. The positions of maxima and minima in the oxide profile do not change for such patterns during planarization.

In the following a simple semi-quantitative model will be presented which illustrates the *relative* performance of various subpad constructions in terms of the length scale of the topography which is being polished. Here "semi-quantitative" means that actual material properties and layer thicknesses are used to calculate anticipated behaviors, but that the situation modeled is too simple to directly use the numerical results for quantitative predictions. The utility of the results is two-fold: The dependence of pad response on length scale is illustrated, and subpad constructions can be ranked relative to their anticipated behavior, considering the length scale of interest. Assume that everything else in the system is the same when comparing different subpads, and that other variables such as the contact compliance of the abrasive, hydrodynamics, carrier film, etc. do not wash out the effect of the subpad. What would one expect changing subpads to do, other effects aside?

In the following developments, the focus is on the reaction of the subpad. The local removal rate depends on both the subpad reaction and the instantaneous pattern density. The subpad reaction depends solely on the deflections of the subpad surface, and is independent of the pattern density. During the formation of doming, the envelope of the oxide surface "up" areas goes from flat to domed, with a maximum range of remaining oxide thickness at the time when the steps are removed in the lowest density area. With the steps removed, the low-density regions become nominally 100% dense. Subsequently, the range decreases until most of the steps have been removed, eliminating the spatial variations in pattern density. Thereafter, the removal rate over a die is driven only by the variation of pressure due to gross topography. For simplicity, the picture presented will be of the final stage of polishing, when all steps have been removed—the dome removal stage. However, the mechanics of the subpad response are unchanged throughout the process of dome formation and removal.

Figure 5 shows a schematic of a long stylus profilometer trace. It is based on actual profilometry. The wafer has been polished so that the steps have been removed. Lower pattern density at the periphery of the dies has led to the formation of doming. The smaller humps are periodic with the same period as the die length. This periodic doming is superimposed on an overall wafer curvature. The wafer is not completely polished; additional oxide could be removed.

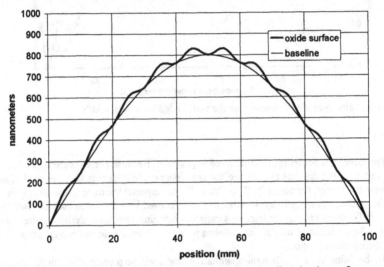

Figure 5. Schematic of periodic die doming superimposed on a baseline that is not flat

At this point in the polish, how does the subpad construction affect the ability to remove the doming? The difference between the pressure at the top of the dome and the bottom of the troughs needs to allow removing the dome in an acceptable time, and before an unacceptable amount of oxide is removed in the low areas. This is a good departure point for the simplified treatment that will be developed.

Sinusoidal Displacement of a Plate on an Elastic Foundation

For simplicity, consider only one-dimensional undulations. The topography can be approximated as sinusoidal. The oxide surface profile looks like a corrugated metal roof. For now consider only the die-scale undulations. The large curvature of the baseline can be treated by superposition. Assume that the mean pressure and the properties of the pad are such that the pad contacts the wafer surface without gaps. Further assume that the contact compliance is low enough that the subpad surface follows the shape of the wafer surface. (The effect of deformation of the abrasive layer would be to reduce the amplitude of subpad deformation while leaving the displacement wavelength the same.) For this simplified treatment, only the effect of normal pressures on the subpad will be considered.

For an imposed sinusoidal displacement, what is the associated pressure distribution on the surface of the subpad? Let the displacement of the stiff subpad layer be

$$w = w_0 + A\cos\left(\frac{2\pi}{l}\right)x \tag{4}$$

where w_0 is the mean displacement, A the amplitude, and l the wavelength. The governing equation for a plate on an elastic foundation in cylindrical bending is

$$D\frac{d^4w}{dx^4} = q(x) - kw \tag{5}$$

where q is the transverse pressure on the plate and x is the position coordinate. Substituting the known displacement into the governing equation, the pressure as a function of position is

$$q(x) = kw_0 + \left[D\left(\frac{2\pi}{l}\right)^4 + k\right]A\cos\left(\frac{2\pi}{l}\right)x \tag{6}$$

As one would expect, the pressure profile looks a lot like the displacement profile. Equation 6 relates the pressure distribution to the subpad properties and the amplitude and wavelength of the displacement.

The kw_0 term corresponds to the mean pressure. The mean displacement is constantly adjusted by the polisher to maintain a constant pressure. For a completely flat surface (A=0), the result is q(x)= kw_0.

The second term relates to the pressure variation due to the topography. This controls the ability to remove material from the high spots faster than the low spots. The term in the brackets (hereafter called "prefactor") scales the amplitude of local pressure variation due to a given dome height, 2A. Planarization of the sinusoidal dome occurs faster for a greater prefactor. The prefactor shows that the length scale of the topography operates on the D, and is a strong function--a fourth power. For small wavelengths, the first term in the brackets (containing D) dominates. For large wavelengths, the second term in the brackets, k, dominates. The actual values of D and k determine when a wavelength is "short" or "long".

The two terms of equation 6 illustrate that globally, there is pressure control by the polisher, but locally, pressure variations causing planarization are determined by the displacements imposed on the pad by the topography. The subpad controls planarization.

Mean Pressure

What difference would mean pressure be expected to make on the pressure response to doming? If the removal rate is proportional to pressure, the answer is "none", since planarization depends only on the second term of equation 6. (If the relationship is non-linear, the effect of mean pressure depends on the exact pressure-removal rate relationship.)

However, the mean pressure also ensures contact of the pad over the entire surface. By setting the pressure at the low point in the profile equal to zero, one can solve for the kw_0 necessary to ensure contact (in this simplified scenario) for a given subpad (i.e., D and k), amplitude A, and wavelength l. The required mean pressure depends linearly on displacement amplitude, but is a strong function of wavelength. As noted above, in the process of the formation and removal of a dome the amplitude of the subpad displacements goes from zero to a maximum, then decreases. The wavelength is constant, and is set by the die size. During polishing, loss of contact with part of the surface, if it occurs, would be expected to only happen during the middle, high amplitude part of the process. Therefore experiments to investigate "bridging" behavior would need to track the whole process.

The term "bridging" as used here does not mean that the pad contacts only the peaks, but that contact is lost at the low points. This simple treatment is no longer applicable to give a sense for the pressure distribution when bridging occurs

Figure 6 shows the dependence of the amplitude of the displacement profile at which bridging is predicted for a given process pressure (5 psi). Various subpad constructions were used for the calculations. All use 0.015, 0.020 or 0.030 inch thick polycarbonate as the top layer. The foams are 0.050 and 0.090 inch thick "standard" foam, and "s50" which is a 0.050 inch thick foam which is about 5 times stiffer than the standard foam. During polishing the wavelength doesn't change. The amplitude rises from zero to some maximum, then decreases. If the pattern-density driven amplitude is above the curve indicated for a given subpad construction, bridging is predicted. As polishing proceeds, the pad begins to make contact with the whole surface--even in the valleys. (For a process pressure of 1 psi, the amplitude scale would be divided by 5. At one-fifth the mean pressure, the amplitude for onset of bridging would be one-fifth of that on the plot.)

As long as there is contact in the valleys, the pressure difference between the peaks and valleys is independent of the process pressure.

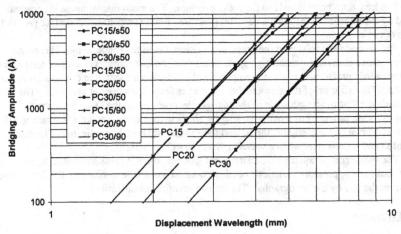

Figure 6. Amplitude of displacement profile for which bridging would occur for 5 psi process pressure (simple model)

This model is too simple to be truly predictive. However, Figure 6 does suggest that for typical process pressures (under 5 psi), bridging will depend only on the polycarbonate thickness, and not the foam layer. This is because the curves for the various subpads collapse onto lines of constant polycarbonate thickness as wavelength decreases. Also, bridging would not generally be expected at the die scale, but is limited to short wavelengths more likely to be associated with features or groups of features within a die.

The Prefactor and Planarization

In this section the relationship of the prefactor to length scale is considered. The prefactor scales the pressure response to the periodic displacement. It is helpful to determine the pressure amplitude for a couple of examples to illustrate the order of the pressure variation for the idealized situation. Take as an example a dome of 1000 Angstrom amplitude (i.e., 2000 Angstrom dome height) over a 10 mm die. The amplitudes of the pressure profile are 0.08 psi for a PC20/90 subpad and 0.37 psi for a PC30/s50 subpad. The pressure variation is an indication of the maximum subpad-driven planarization capability for these subpads in this situation. Note that the pressure amplitudes will fall off directly with the amplitude of the displacement profile. Deformation of the abrasive and contact compliance of the polycarbonate layer will attenuate the displacements imposed on the subpad. These pressures can be considered to be maximum values for the ideal situation where those factors are negligible.

The most helpful way to use the simple model is to rank subpad constructions relative to each other. For this, only the prefactors need be considered. Given a wavelength, the prefactor can be calculated for various candidate constructions. Then the higher the prefactor, the more quickly the pad will planarize a dome of a given amplitude. Figure 7 is a plot of the prefactor as a function of displacement wavelength for the subpad constructions of Figure 6.

At short wavelengths, the curves collapse on lines of constant D--of like polycarbonate thickness. At long wavelengths, the curves collapse on lines of constant k--of like foam thickness and modulus.

Generally speaking, the left third of the plot relates to sub-die scale features, the middle third relates to the die scale, and the right third relates to the wafer scale.

It is important at this point to note that a higher prefactor is not always better. Refer back to Figure 5, depicting two superimposed wavelengths. One would not want to remove the long wavelength curvature. The objective is actually to selectively remove small deviations from a gradually curving baseline. Selectivity and sensitivity as a function of scale are key to accomplishing this. The longer-wavelength topography will have a lower prefactor for a given construction, but because the amplitude is greater than for the shorter-wavelength undulations, the associated pressure difference can be large relative to that due to the short wavelength features. The best subpad would have a suitably high prefactor at the die scale, and a sufficiently low prefactor at the scale of wafer bowing, for example.

Wafer uniformity generally depends on the long scale, so it is enhanced by having softer, thicker foam (i.e., lower k). However, uniformity relating to the edge of the wafer depends on the characteristic length, as seen above. For a given polycarbonate thickness, reducing the foundation modulus increases the characteristic length, and thus the size of the non-uniform edge region.

At the short, sub-die scale, the prefactor (and planarization capability) depends only on the polycarbonate thickness.

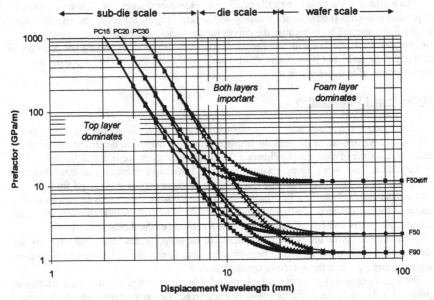

Figure 7. Prefactor as a function of displacement wavelength and subpad construction. The foam is listed at the right border; the polycarbonate layer thickness is listed at the top border.

At the long and short length scales, the relative subpad behaviors are clear. It is at the die or intermediate scale that the details of the interaction of length scale and composite subpad construction become important.

First, note that there are crossover points in the plot. Those are points identifying wavelengths or scales at which pads will have the same response. For example, the curves for PC20/s50, PC15/s50, PC30/50, and PC30/90 cross at a wavelength a little over 10 mm. One would expect, then, that for a die of that size one would see a negligible effect on doming when comparing the performance of those subpads. For small features the PC30/50 or PC30/90 would be the best, and would give similar performance. For uniformity (long wavelength), the PC30/90 would be the best because of its lower prefactor. In general, the PC30/90 would be expected to give the best performance overall, but with respect to die doming, it would be no different than the three others. Another crossover point can be seen at about 19 mm. There the curves for PC20/50, PC15/50, and PC30/90 are close.

The expectation of crossovers is important for understanding experimental subpad comparisons. *Whether or not changing the subpad will improve doming (or intermediate length scale) performance depends on the specific scale imposed by the wafer topography.* There will be scales that will be insensitive to some changes in subpad construction.

Note that the sensitivity of the prefactor to wavelength varies, i.e., that the slope is especially sensitive at the die scale and that the range of wavelengths over which the slope changes is somewhat different for each subpad. This is important, since one wants to be able to choose the subpad to preferentially remove humps below a certain wavelength. Of all the possible subpads of this type, what subpad should be chosen, given a target wavelength? That wavelength would be set by the die size or perhaps some other large feature. One can plot the prefactor curves for all subpad constructions that have equivalent performance at the target wavelength. Figure 8 shows the curves for subpads that have a prefactor of 50 GPa/m at a displacement wavelength of 10 mm. The legend lists the foundation modulus for the subpads.

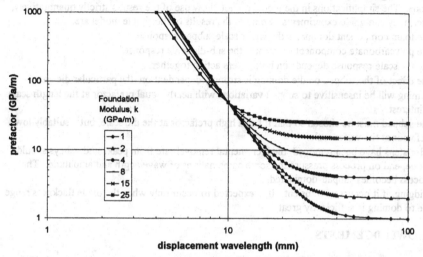

Figure 8. Prefactor curves for subpads having a prefactor of 50 GPa/m at a displacement wavelength of 10 mm.

The main result is that for a family of subpads like this maximum sensitivity at the target wavelength (i.e., slope) belongs to the pad with the lowest k. It will have the highest planarization ability at wavelengths below the target wavelength, and the lowest prefactor for wavelengths higher than the target.

On the other hand, if there were features of similar but varying scales, a pad with a flatter slope over the scales of interest would be preferred.

CONCLUSIONS

For structured abrasive CMP the pressure distribution at the process surface depends largely on the subpad response. That response depends on the length scale of the loading. Contact mechanics and plate bending mechanics have been used to illustrate the nature of the effects of length scale. This understanding is meant to inform experiments to optimize the CMP process for particular situations.

The effect of subpad construction is expected to be greatest in the multi-millimeter length scales. The thickness of the abrasive layer attenuates the ability of the subpad to respond to details of loading at the process surface.

Single-layer and two-layer subpads respond in markedly different ways to point loading. The single-layer subpad response is dominated by contact mechanics, while the two-layer subpad response is dominated by plate bending. The influence of loading at a point is over a much larger area for the two-layer pad.

A simple semi-quantitative treatment based on the model of a plate on an elastic foundation was presented. It was used to illustrate the relative performance of various subpad constructions in terms of the length scale of the topography that is being polished. The context was die doming, but the results are relevant to other periodic multi-millimeter length scale load situations. The simplifications in the model do not allow use of the results strictly quantitatively; however, they can guide experiments. Some of the results of the simple model are:

- The foam component dominates the wafer scale subpad response.
- The polycarbonate component dominates the sub-die scale response.
- The die scale response depends on both layers acting together.
- The effect of the subpad on die doming is strongly dependent on the particular die size.
- Doming will be insensitive to subpad variations with nearly equal prefactor at the length scale of interest.
- The subpad must be chosen so that it has a high prefactor at the die scale, but a suitably low prefactor at the wafer scale.
- After steps have been removed, the incremental removal rate from peaks and valleys should not depend on process pressure for some combinations of wavelength and amplitude. The process becomes subpad-dominated.
- Bridging, if it occurs, is transient. It is expected to occur only when the oxide thickness range due to doming is sufficiently great.

ACKNOWLEDGEMENTS

The author gratefully acknowledges the kind help of Dr. Robert Jennings in generating the finite element results of Figure 3, and for review of the manuscript.

REFERENCES

1. E. F. Funkenbusch, "A Slurry-Free CMP Technique for Oxide Planarization ", Proc. CMP Technology for ULSI Interconnection Symp. (Semicon West) San Francisco, July 14, 1998.

2. M. Fayolle, J. Lugand, F. Weimar, and W. Bruxvoort, "Evaluation of a New Slurry-Free CMP Technique for Oxide Planarization", Proc. Third Int. Chemical-Mechanical Planarization for ULSI Multilevel Interconnection Conf. (CMP-MIC) Santa Clara, CA, February 19-20, 1998.

3. D. P. Goetz, "The Effect of Subpad Construction on Pattern Density Effects for Slurry-free CMP", Proc. Fourth International Chemical-Mechanical Planarization for ULSI Multilevel Interconnection Conf. (CMP-MIC) Santa Clara, CA, February 11-12, 1999.

4. K. L. Johnson, *Contact Mechanics* (Cambridge University Press, Cambridge, 1985), p. 50-51.

5. S. Timoshenko and S. Woinowsky-Krieger, *Theory of Plates and Shells*, 2nd ed. (McGraw-Hill, New York, 1959), p. 270.

6. Ibid., p. 267

7. D. P. Goetz, "3M Subpad Mechanics: A Framework for Characterization", Proc. Northern California Chapter of the AVS CMP Users Group, San Jose, CA, September 2, 1998.

A Novel Retaining Ring in Advanced Polishing Head Design
for Significantly Improved CMP Performance

Thomas H. Osterheld, Steve Zuniga, Sidney Huey, Peter McKeever, Chad Garretson, Ben Bonner, Doyle Bennett, Raymond R. Jin
Applied Materials, 3111 Coronado Drive, M/S 1510, Santa Clara, CA 95054

ABSTRACT

This paper reports a technological advancement in developing and implementing a novel retaining ring of advanced edge performance (AEP™ ring) for an advanced polishing head design. The AEP ring has been successfully used for significantly improved CMP performance in different CMP applications: oxide (PMD and ILD), shallow trench isolation (STI), polysilicon, metal (W and Cu), silicon-on-insulator (SOI), and silicon CMP. Robust processes have been developed using AEP ring along with many hardware upgrades for each application with extended runs to meet requirements of advanced IC device fabrication.

INTRODUCTION

The challenge in polishing head design is to assure required CMP process performance (low non-uniformity, high removal rate, low defects, good head-to-head matching, and low-pressure/ high-speed capability), short qualification time, and low cost of ownership. Conventional polishing heads coupled with polyurethane-based polishing pads have a limitation in achieving a low non-uniformity due to the pad deformation at the edge of the wafer during polishing, as manifested by a slow or fast edge polishing rate. In this work, an advanced polishing head (Titan Head™) design has been studied by utilizing different retaining rings. The AEP retaining ring has been developed and implemented to shorten polishing head qualification time and to achieve the CMP process performance required for manufacturing advanced IC devices (e.g., sub-0.25μm and/or mixed signal devices).

EXPERIMENTAL

A 4-head and 3-platen CMP tool (Mirra® CMP system) was used with 8" wafers throughout this work. Different retaining ring designs coupled with a Titan Head design were investigated for improved performance. The Titan Head design features a flexible membrane applying a uniform pressure to the backside of the wafer. The advanced polishing head is equipped with a retaining ring to prevent wafer slippage during polishing. A more important function of the retaining ring in a Titan Head design is to modulate the polishing removal rate near the edge of the wafer for a low within wafer non-uniformity (WIWNU) at reduced edge exclusions. This is achieved by independently controlled pressure applied onto the ring. Two different types of retaining ring materials were used in this work. The first type is a polymeric material requiring lapping. The second type used in different ring designs including the AEP design is the layered materials. The retaining ring made of the first type of material is called P ring in this paper. The new-generation retaining ring (AEP ring) requires no individual modification for fitting, no lapping and no adjustments.

Stacked IC1000/Suba IV pads (Rodel) were used for polishing and a Politex pad (Rodel) was used when buff is applied. For pad conditioning, an improved diamond disk manufactured by using a proprietary high temperature Ni-Cr brazing process was used in conjunction with an advanced conditioner head assembly design. [1] Cabot's SS-12 slurry was used as an oxide slurry. Consumables used in W and polysilicon CMP were described in previous work. [2, 3].

RESULTS AND DISCUSION

Impact of Retaining Ring Design Types on WIWNU

The retaining ring design has a significant impact on WIWNU due to the ring-pad-wafer-membrane interaction (Fig. 1). In a Titan Head design, an independently-controlled pressure (P_1) is applied onto the retaining ring to absorb pad deformation around the wafer edge during polishing. Meanwhile, a uniform pressure (P_2) is applied by a flexible membrane over the wafer backside to achieve a desirable polishing rate. To optimize WIWNU, the retaining rings of different designs were evaluated. The results are summarized in Table 1. The results indicate that an AEP ring design is critical to ensure low WIWNU. The AEP ring design can more effectively control pad deformation during wafer polishing as compared to the conventional P ring. As a results, better edge performance of an AEP ring was achieved than a P ring as shown in wafer scan (Fig. 2).

Impact of Pressure and Speed on Removal Rate and Non-Uniformity

Designed experiments were conducted to optimize the processes for high removal rate and low non-uniformity. The impact of retaining ring pressure and platen speed on removal rate and WIWNU is shown in a contour map (Fig. 3).

The results indicate that WIWNU can be significantly improved by increasing retaining ring pressure in the range studied which is made possible by Titan Head design. At a high retaining ring pressure, WIWNU can also be improved by increasing platen speed at a constant head/platen speed difference. However, when retaining ring pressure is lowered, the trend is different. WIWNU can be improved by decreasing platen speed. The results also indicate that removal rate is proportional to platen speed and insensitive to retaining ring pressure at a constant membrane pressure. Later studies reveal that the contour map is more complicated when speed and pressure are extended to outside of the range studied in this work. [4]

Removal Rate and WIWNU in Oxide CMP Extended Runs: AEP vs. P Ring

An extended run was conducted using AEP rings for 3 heads and a P ring for the fourth head with thermal oxide wafers. The results are summarized in Table 2 and Figs. 4 - 5. Table 2 shows that AEP rings improved wafer-to-wafer non-uniformity (WTWNU) (1.6%) as compared to the P ring (2.5%) in this extended run (called Run #2 in Table 2). The performance of P ring in this extended run (Run #2 in Table 2) is similar to another extended run using P rings on all four heads (Run #1 in Table 2). For comparison, AEP performance from the third extended run using

Table 1. WIWNU as a function of retaining ring design types.

Retaining Ring Design Types	WIWNU at 5mm EE
AEP	1.0 to 3.0%
Alter_A	2.4 to 3.0%
Alter_B	3.1 to 4.0%
Alter_C	8 to 14%

Table 2. Removal rate and WIWNU with AEP ring as compared to P ring in extended runs. (Run #2 presented in Fig. 4. A different process at higher pressure used in Run #3).

Type of Retaining Ring	P	P	AEP	AEP
Extended Run #	Run #1	Run #2	Run #2	Run #3
Process Performance				
Thermal Oxide Removal Rate (A/min)	2,900	3,000	3,000	4,050
PETEOS Removal Rate (A/min)	3,600	3,750	3,750	5,050
WIWNU (5mmEE-1s)	3.2%	2.7%	2.9%	2.0%
WTWNU (5mmEE-1s)	2.1%	2.5%	1.6%	2.6%

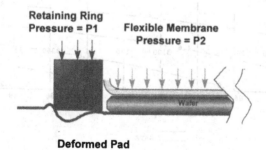

Fig. 1. Schematic diagram of retaining-ring/pad/wafer/membrane interaction.

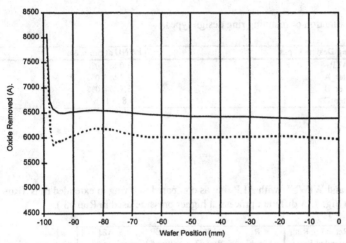

Fig. 2. Wafer scan from AEP (solid line) and P (dotted line) retaining rings using an optimized process.

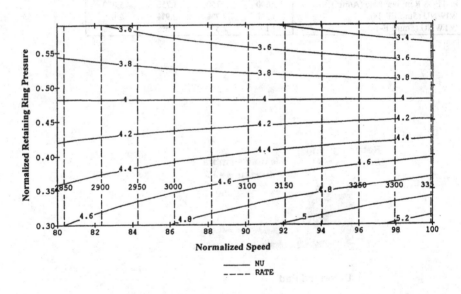

Fig. 3. Oxide removal rate and WIWNU (3 mm EE)contour map plotted against normalized platen speed and normalized retaining ring pressure at a constant membrane pressure and a constant head/platen speed difference.

a higher pressure process and AEP rings on all four heads is also listed in Table 2 as Run #3. The results indicate that WIWNU and removal rate using the AEP ring without lapping are stable and comparable to those with the lapped P ring using the same process (pressure). To achieve acceptable CMP performance at <5 mm edge exclusion, a P ring is always lapped prior to CMP operation. An AEP ring does not require lapping. An AEP ring design ensures precise and uniform pressure control on a polishing pad around the wafer edge.

The AEP ring has been implemented in IC fabrication. An 800-wafer extended run was conducted over three days, the results confirmed the stability of removal rate and WIWNU achieved by an AEP ring (Fig. 6).

Defect Performance Using AEP Ring

After the lapping operation using a conventional lapping tool, macro-scratches and embedded particles from abrasive lapping materials may appear at the working surface of the P ring. The embedded particles could release during subsequent polishing steps and thus scratch the wafer. The AEP ring does not require lapping and thus eliminates the possibility of scratching by embedded abrasive lapping materials and thus provides reduced micro-scratch and defect counts on the wafer surface. Typical defect performance over multiple days using 2-platen polish with the third platen rinse is shown in Fig. 7. The defect counts (>0.20μm, SS6420) are less than 45 (UCL) with average count of 18. The total defect and micro-scratch counts (>0.25μm, SS6200) are less than 88 (UCL) with average count of 33. Similar defect performance was achieved using 3-platen IC1000 polish with the third platen rinse for improved throughput. The average defect count (>0.20μm, SS6420) is 18 and the average total defect and micro-scratch count (>0.25μm, SS6200) is 39 over a 1200-wafer extended run using a 3-platen IC1000 polish.

Customer Performance: AEP vs. P Ring

The AEP ring performance was monitored in a high volume production fab. A test wafer was measured for removal rate, WIWNU and defect every 25 wafers. Over a one-month run, AEP rings significantly improved WIWNU. As compared to P rings, AEP rings reduced average WIWNU by more than 40% to 3% at 5 mm edge exclusion. AEP rings also eliminated all the out-of-control points. AEP rings on two different Mirra tools also showed improvement in defect counts. The improvement by AEP rings as compared to P rings is attributed to the AEP ring design and elimination of lapping. As required, P rings were lapped off-site by different operators on a relatively low-tech lapping tool after each head re-build. Quality of lapped P rings used in a fab is less consistent than that of AEP rings without lapping.

Removal Rate, WIWNU and Defects in a W CMP Extended Run

The AEP ring has been successfully applied to W CMP. [2] A 1600-wafer W CMP extended run has been conducted using an AEP ring. As seen in the oxide CMP application, a stable removal rate (4046 Å/min.) and low and stable non-uniformity (average WIWNU 2.5%, WTWNU 4.1%) was achieved at 5mm edge exclusion (Fig. 8). Defect counts on PETEOS wafers (>0.25μm, SS6200) in the W CMP extended run were kept below 100 for the process with oxide buff.

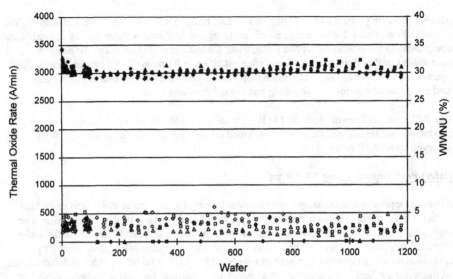

Fig. 4. An extended run using AEP rings for 3 heads and a P ring (square symbol) for the fourth head with thermal oxide wafers at 5 mm edge exclusion.

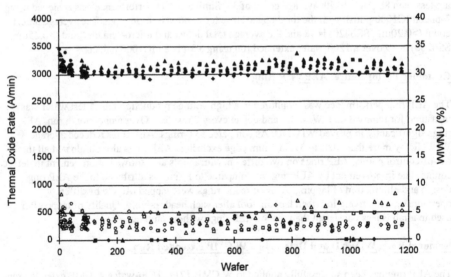

Fig. 5. An extended run using AEP rings for 3 heads and a P ring (square symbol) for the fourth head with thermal oxide wafers at 3 mm edge exclusion.

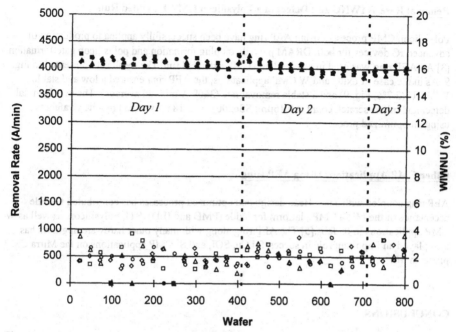

Fig. 6. An extended run over 3 days using AEP ring with oxide wafers at 5 mm edge exclusion.
Avg. removal rate = 4053 Å/min. Avg. WIWNU = 2.0%. WTWNU = 2.6%.

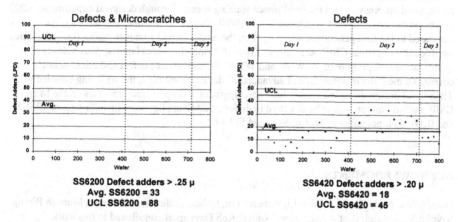

Fig. 7. Defect and micro-scratch counts in an extended run over 3 days with on oxide wafers
using AEP ring and a 2 platen polish process.

Removal Rate WIWNU and Defect in a Polysilicon CMP Extended Run

Polysilicon CMP processes using AEP ring have been successfully applied to production of advanced IC devices for both DRAM polysilicon plug formation and polysilicon gate formation. [3] An AEP ring was used in a polysilicon CMP extended run. The results are presented in Fig. 9. As in the case of oxide and W CMP applications, the AEP ring ensured a low and stable WIWNU (2.7%, Fig. 9) and a stable removal rate (3609 Å/min. on average). The average total defect and micro-scratch count (>0.16µm, SS6200) from 18 polysilicon monitor wafers is 25 using an optimized process.

Other CMP Applications Using AEP Ring

AEP ring coupled with Titan Head design and optimized processes has contributed to wide acceptance of the Mirra CMP platform for oxide (PMD and ILD), STI, polysilicon, as well as W CMP applications in IC fabs.[5] The AEP ring along with many other hardware upgrades has also played an important role in successful Cu, SOI, and Si CMP applications on the Mirra CMP platform. [6]

CONCLUSIONS

The retaining ring used in an advanced polishing head design improves WIWNU by modulating pad deformation and controlling the removal rate around the wafer edge. WIWNU and removal rate depend on pressure and the head/platen rotating speed. Through designed experiments, CMP processes can be optimized to achieve low WIWNU, high removal rate, and stability. As compared to the first generation retaining ring, the retaining ring of novel design (i.e., AEP ring) significantly improves CMP performance. The AEP ring eliminates lapping, thus reduces qualification time, improves WIWNU and WTWNU, and reduces defect and micro-scratch counts. To meet the requirements of advanced IC device fabrication, the AEP ring coupled with Titan Head design along with many hardware upgrades and optimized processes on the Mirra CMP platform are proved to be effective in different CMP applications, including oxide (PMD and ILD), STI, polysilicon, W, Cu, SOI, and Si CMP.

ACKNOWLEDGMENTS

Hung Chen, Andy Barda, Shijian Li, Robert Lum, Robert Tolles, Fritz Redeker, Manush Birang, Savitha Nanjangud, Bret Adams, Greg Amico, Rob Davenport contributed to this work.

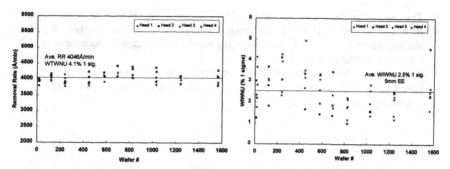

Fig. 8. An extended run using AEP ring with W wafers at 5 mm edge exclusion.

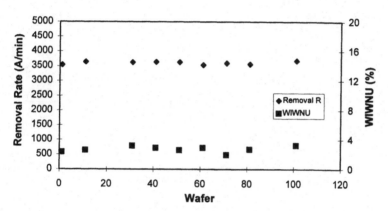

Fig. 9. An extended run using AEP ring with polysilicon wafers at 5 mm edge exclusion.

REFERENCES

1. Raymond Jin, Gopal Prabhu, Steve Mear, Sidney Huey, Robert Tolles, Fritz Redeker, "Significant improvement of disk/pad life and polishing performance by using a new pad conditioning disk", proceedings of 1998 VMIC conference, pp. 527-529.
2. Robert Lum, Sourabh Mishra, Fritz Redeker, Brian Brown, Kapila Wijekoon, Ronald Lin, Savitha Nanjangud, "Defects characterization and productivity optimization of tungsten CMP process", presented at 1999 SEMICON Korea, Seoul, Korea, February 22-24, 1999.
3. Raymond Jin, Shijian Li, Simon Fang, Fritz Redeker, "Proven practice and future application of polysilicon CMP in IC fabrication", proceedings of 1998 SEMICON Taiwan, pp. 263-274.

4. Thomas H. Osterheld, Chad Garretson, Peter McKeever, Ben Bonner, to be published.
5. Raymond Jin, Jeffrey David, Bob Abbassi, Thomas Osterheld, Fritz Redeker, "A proven shallow trench isolation (STI) solution using novel CMP concepts", proceedings of 1999 CMP-MIC conference, pp. 314-321.
6. Raymond Jin, "New generation CMP equipment and its impact on IC devices", 1998. proceedings of 5th International Conference on Solid-State and Integrated-Circuit Technology, pp. 116-119.

CMP FUNDAMENTALS AND CHALLENGES

Michael R. Oliver
Rodel, Inc., Phoenix, AZ

ABSTRACT

Chemical Mechanical Polishing (CMP) as a semiconductor polishing technology has grown dramatically during the past decade. It has been a key enabling technology, facilitating the development of high density multilevel interconnects. Its widespread application has exceeded the growth of the scientific understanding.

Models for silicon dioxide polishing mechanisms have built upon the work originally done for glass polishing. Recent work has augmented our insight, but our understanding is far from complete. A quantitative picture of the basic interaction mechanisms for silicon dioxide polishing does not yet exist. The models for metal polishing are now an active area of investigation. Recent work has demonstrated that more work is need to adequate explain the CMP of metals. The metal CMP mechanisms appear to be substantially more complex than originally assumed.

The challenges for CMP have been to improve the performance of the technology, and increasingly this needs a foundation of scientific understanding to achieve the needed gains. We can expect that, as the scientific foundation grows, it will contribute significantly to manufacturing improvements as has been the case in other areas of semiconductor technology.

INTRODUCTION

Chemical Mechanical Polishing (CMP) has emerged over the last decade as an enabling technology within the semiconductor industry. It has allowed the widespread application of high density multilevel interconnects, and has also found use in applications such as shallow trench isolation (STI). There will soon be almost 2000 polishing machines being used for CMP world wide.

The technology itself has advanced tremendously as substantial improvements in capability have led to greatly improved performance with increasingly tight process control. Many semiconductor processes used in manufacturing now have over ten CMP steps within them; this revolution in semiconductor technology has occurred because of demand pull. CMP based processes can be more powerful than those processes without CMP.

This advance of CMP technology has naturally been directed toward application to the semiconductor industry. The large majority of the "research and development" effort has focussed on near term process improvements, in hardware, consumables and in the CMP process itself. In all of these areas the current processes in use are markedly different from those in use 8-10 years ago. Currently, the effort focussed on further advances in all three areas is larger than it ever has been.

The basic physical and chemical mechanisms of CMP are, however, poorly understood. The polishing process as used in semiconductor applications is the sum of a very large number of polishing interactions, with a range of conditions determined by the tools used. The sum of these individual interactions creates the polished surface. The specific conditions of these individual interactions are not well understood and it is quite difficult to set up experiments and measure local interactions which correspond meaningfully to those used in practice. This in itself has

73

limited modeling and development of mechanisms. It has also led to various alternative models with assumptions used where specific information is lacking [1].

This lack of insight to individual polishing mechanisms will be a growing limitation as the gains obtained from the application technology are more completely exercised and further first principles understanding will be required for focussing development work based on clearer understanding of fundamental interactions. A brief summary of the current understanding of mechanisms can show this limitation.

BASIC MECHANISMS

For polishing to proceed and create a nearly planar surface, interactions and material removal must occur in or near a plane. For a pad and slurry system as used in CMP, the pad is in contact with abrasive particles which are in contact with the material covering the wafer which is to be removed. This is qualitatively pictured in Figure 1. Here, the particle is idealized to be spherical, which is usually not the case, and the asperity structure of the pad is crudely simplified, except to show that the tops of the asperities are flattened somewhat, which is often observed.

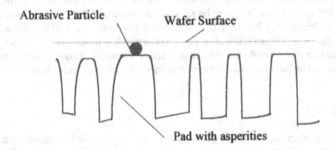

Figure 1. Schematic showing the three basic components of CMP interactions: wafer surface, abrasive particle and pad with asperities.

Silicon Dioxide – Abrasive Mechanism

There have been recently several papers addressing the polish rate variations for several materials with pressure and velocity, and several different relationships have been obtained, depending upon the assumptions of pad-abrasive particle-wafer surface interaction behavior[1]. These different results have been obtained because there is no quantitative, detailed interaction model which has been confirmed by experiment and is predictive. The difficulty in achieving this experimental understanding is high, and there is not a good body of knowledge to build upon.

Prior to the advent of CMP in the semiconductor industry, the scientific literature on polishing in general was sparse. The major share of polishing technology was focussed on glass polishing for a large variety of applications, and many patterns of behavior had been observed and documented. In 1990, Cook[2] summarized the knowledge of glass polishing characteristics known to that time. He focussed on the glass-abrasive interaction, and described observed relationships between polish rate and the nature of the abrasive. As an element of one potential mechanism, he discussed the model of the Hertzian indenter, and several potential contributing mechanisms for material removal. The Hertzian indenter particle being forced into the surface to be polished is pictured in Figure 2. Cook reviewed possible effects that could contribute to

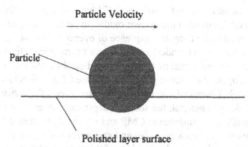

Figure 2. Model of Hertz indenter, with abrasive particle indenting into the polished layer as it travels with a velocity.

material removal from the glass surface with the indenter, and did not find good quantitative agreement with the polishing data. Cook and Brown [3] had earlier proposed the concept of a "chemical tooth" to account for material removed in the polishing regime. Cook reviewed possible chemical interactions and noted some correlations of various metal oxide properties with polishing performance. With the experimental data available, this summary provided some directions for future studies, and he noted specific studies that would aid in this understanding.

Quantitative experimental confirmation of different possible mechanisms with the model would be very difficult because of the idealizing simplifications that are made. The indenter model, however, has provided an available starting point for CMP studies as glass polishing technology is similar in many ways to semiconductor silicon dioxide polishing. As noted, several CMP investigators who have used this model[1] to analyze published data have made differing assumptions about the details of the model.

It is interesting to note that, for the abrasives used to polish glass, silica abrasives are so ineffective as not to be considered. This is in contrast to silicon dioxide polishing in CMP where silica abrasive is used almost exclusively as present.

Another general type of model that has been presented is the shear erosion polishing approach as described by Runnels[4]. This model omits details of the abrasive interaction, and thus does not address the specific polishing mechanism. In the absence of specific interaction mechanisms, similar approaches [5] have been used to predict polishing results for arbitrary wafer structures.

Metals

Metal polishing mechanisms appear to be considerably different from those for silica polishing. The first mechanism specifically proposed for metal CMP was that of Kaufmann et al. who proposed a relatively straightforward two step mechanism, focussing on tungsten polish [6]. The first step is the oxidation of the metal surface by an oxidant in the slurry, and the second step is the abrasion removal of the oxide from the metal surface by the abrasive in the slurry. This model accounts for the observed experimental behavior in tungsten polish of an oxidant in the slurry, as well as the abrasive. For this model to proceed, the initial oxidation of the tungsten surface must be more rapid than the oxidation of a surface covered with a thick layer of tungsten oxide. Generally, the model agrees well with a large number of experimental observations on polishing processes. In addition, the model predicts reasonably well the polish rate variations as functions of oxidants and abrasives in the slurry.

Recently detailed analyses by Stein et al.[7] have determined that the observed polish rates for tungsten are inconsistent with the known oxidation rates of tungsten in specific oxidants. Closer analysis shows a more complex sequence of events. However, at this point, the intermediate events are not well understood and await further investigation.

There has been a great deal of effort focussed on the understanding of the interaction on copper polishing. Copper damascene processes require CMP. Steigerwald, Murarka and Gutmann in their book, *Chemical Mechanical Planarization of Electronic Materials [8]* indeed review CMP technology in general, but focus in depth on copper issues. Copper behaves significantly differently from tungsten in CMP and it can be polished at low pH under non-passivating conditions. The book provides an excellent summary up to 1995. Cook and Wu [9] provide more experimental data and an update of the status to 1998.

This wealth of copper CMP data has not narrowed the focus of developing an abrasive particle-copper interaction model, but actually expanded it. Copper can be polished with various abrasives, in the presence of various suppressants and with a wide range of oxidants over a wide pH range. Because of the very rigorous demands of copper damascene CMP, and the incomplete current understanding, the research effort focussed on copper is still growing.

Pad-Polishing Abrasive-Wafer Interaction

The third element of the interaction, is the pad as shown in Fig. 1. The purpose of the pad is to provide uniform pressure to the asperities in a narrow range in the direction normal to the wafer. In this way, the material removal occurs so that the polished material surface evolves toward a planar one. There are a number of additional properties of the pad that are important and have been empirically optimized. These properties include the pad hardness, its ability to be conditioned, its density, its moduli, as well as several other physical and chemical properties.

Another important part of the overall system that has to be considered is the slurry layer between the wafer and the pad. This slurry fills the space and transports the abrasive particles, as well as mediate the chemical-mechanical interactions between the abrasive and the polished material, especially in metal polish systems. Several workers have looked at the slurry layer, especially in simplified systems, have found slurry thicknesses, experimentally in a range near 30-40 microns[10,11]. This is roughly the measured heights of pad asperities in standard semiconductor CMP applications[12]. The actual volume of this incompressible layer per unit area clearly plays a role in how the asperities interact with the wafer and pad, but has not yet been closely examined.

In all of the above discussion, the focus has been entirely on the local conditions where the point on the wafer, an abrasive particle and the pad meet. In practice, the system is substantially more complicated, with much effort directed toward producing a clean, uniform and manufacturable process. Thus the polishing machine usually produces a complex wafer motion, including wafer rotation, which averages polish conditions which vary with position across the wafer. In addition, the pad and the asperities created also have position dependence for which complex wafer paths and conditioning procedures are used. These useful and necessary techniques enhance the performance of the technology but also obscure the behavior of individual polishing events.

CHALLENGE OF UNDERSTANDING BASIC MECHANISMS

The difficulty of extracting information about basic mechanisms from CMP technology as used in manufacturing is, as noted above, extremely difficult. The observed effects are the sum of a very large number of polishing events over a range of operating conditions, which are effectively averaged.

In addition, for modeling both the interactions between the abrasive particle and the polished material, and the interactions between the polished material, abrasive particle and the pad asperities, a large number of materials and properties need to be taken into account. It has been experimentally determined that CMP behavior can be modified significantly by the mechanical and chemical properties of :

A.   the film being polished
B.   the type, morphology and size of the abrasive particle
C.   the slurry liquid composition, including viscosity, surface energy, ionicity, and compostion including other chemicals
D.   the pad surface structure, the pad chemistry and morphology, the pad chemical properties, and the pad surface energy interactions with the slurry
E.   the normal force on the system and the relative velocity of the wafer surface relative to the pad
F.   the amount of slurry contained in the volume under consideration.

This list is not complete, but does indicate that the system considered is complex, and that identifying and controlling the significant variables is certainly a challenge.

Because of the complexity of the system, it is highly desirable to characterize the behavior of individual interactions while other components are held fixed, or nearly so. The key interaction that is the most important to the whole process is the interaction of the individual abrasive particle ith the polished surface. This basic interaction, repeated on a grand scale, is what leads to a polished surface. To measure such a particle-surface interaction and its material removal effects will constitute a major advance.

One set of tools that can be a major advance is AFM technology. This technology, and its variations, are providing many new ways to measure quantitatively very small individual particles interacting with surfaces under highly controlled conditions. Several laboratories, including ours and Sandia's, are exploring slurry particle interactions with different surfaces in different liquid.

- **Attached particles to AFM cantilever**
- **Experiments can be performed using tip in various solutions**
- **Adhesive force**
  - drive tip against surface and retract
- **Friction force**
  - scan tip over surface and measure horizontal deflection
- **This is the closest we can come to measuring the particle-surface interaction during CMP in a direct experiment**

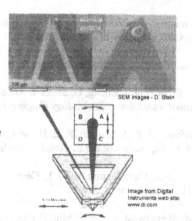

Figure 3. Schematic drawing of AFM and SEM of particle attached to cantilever. Courtesy of David Stein, and Digital Instruments.

ambients. An AFM cantilever probe and SEM images of an alumina particle on it are shown in Figure 3. Attractive and repulsive forces from surfaces can be quantitatively measured as a function of applied force and particle distance from the surface. At present, several slurry particles are put on the cantilever at one time, and this limits somewhat the understanding, but it is reasonable to expect that soon individual particle interactions will be measured

CONCLUSIONS

AFM and other techniques have just begun to be employed to meet the challenge of obtaining understanding of individual polishing interactions. Analogously to other semiconductor technologies, such as plasma etch, this insight will lead to a second wave of advances, based on development directed by known physical and chemical laws.

The next few years could very well be the most exciting of the CMP industry as the technology evolves from the "art" phase to a "science" phase. There are a wealth of exciting directions to work toward to provide meaningful advances to the technology. It is highly probable these new discoveries will trigger alternate approaches to current technology, and continue to fuel the growth of the overall semiconductor industry.

ACKNOWLEDGMENTS

The author would like to acknowledge the contributions of Lee Cook, David Evans, Dale Hetherington, and David Stein, for valuable discussions and information.

REFERENCES

1. B. Zhao and F. G. Shi, *Proceedings of Chemical-Mechanical Planarization for ULSI Multilevel Interconnect Conference*, p. 13, 1999, and references contained therein.

2. L. M. Cook, Journal of Non-Crystalline Solids **120**, p. 152 (1990).

3. N. Brown and L. M. Cook, Paper TuB-A4, Tech. Digest, Topical Meeting on the Science of Polishing, Optical Society of America, April 17, 1984.

4. S. Runnels, Journal of Electronic Materials **25**, p. 1574 (1996).

5. D. Ouma, D. Boning, J. Chung, G. Shinn, L. Olsen, and J. Clark, *Proceedings of the International Interconnect Technology Conference, p. 67, 1998*

6. F. B. Kaufman, D. B. Thompson, R. E. Broadie, M. A. Jaso, W. L. Guthrie, D. J. Pearson, and M. B. Small, Journal of the Electrochemical Society **138**, p. 3460 (1991).

7. D. J. Stein, D. L. Hetherington, and J. L. Cecchi, Journal of the Electrochemical Society, accepted to be published in 1999.

8. Joseph M. Steigerwald, Shyam P. Murarka, and Ronald J. Gutmann, *Chemical Mechanical Planarization of Microelectronic Materials*, Wiley, New York, 1997.

9. L. M. Cook and G. W. Wu, *Proceedings of the Chemical Mechanical Planarization for ULSI Multilevel Interconnect Conference*, p. 150, 1998.

10. J. A. Levert, S. M. Mess, R. F. Salant, S. Danyluk, and A. R. Baker, Tribology Transactions **41**, p. 593 (1998).

11. J. Coppeta, C. Rogers, L. C. Racz, A. Philipossian, and F. B. Kaufmann, *Proceedings of the Chemical Mechanical Planaraization for ULSI Multilevel Interconnect Conference*, p. 37, 1999.

12. A. R. Baker, *Proceedings of Chemical Mechanical Planarization for ULSI Multilevel Interconnect Conference*, p. 339, 1997.

Part III

Metal Polishing—W and Al

DEVELOPMENT OF A ROBUST KIO3 TUNGSTEN CMP PROCESS

ALBERT H. LIU *, RANDY SOLIS **, JOHN GIVENS **
*CMP/New Technology Engineering, VLSI Technology, Inc. San Antonio, TX 78251,
albert.liu@vlsi.com
**CMP/New Technology Engineering, VLSI Technology, Inc. San Antonio, TX 78251.

ABSTRACT

A production worthy, Tungsten Chemical Mechanical Polish (CMP) process using a commercially available KIO_3 slurry was developed, characterized, and tested for sub-0.35µm multilevel interconnect fabrication. The effects of pre-tungsten CMP process on tungsten polish are reported in detail. A head-to-head comparison of the optimized KIO_3 process with the standard $Fe(NO_3)_3$ process is described. Critical CMP tool parameters (process and hardware) were flexed using statistically valid experimental designs. The advantages and disadvantages of a post tungsten polish, oxide buff, are discussed. Across-wafer non-uniformity, specifically the enhanced polish rate of tungsten at the wafer edge, was significantly reduced with the optimized process parameters and hardware setup. Also, an automated endpoint system was utilized and a set of robust endpoint algorithms were developed to minimize the amount of oxide loss during tungsten CMP processing. Finally, the positive effects of the optimized KIO_3 tungsten CMP process on interconnect integration and die yield are reported.

INTRODUCTION

As circuit density increases in VLSI technology, Chemical Mechanical Polishing (CMP) of dielectric material has been widely utilized in the semiconductor manufacturing flow for planarization purpose. However, results from shrinking device geometry have also shown that Al step coverage in contacts and vias limits interconnect design [1]. To ensure sufficient Al step coverage, sloped or wet/dry processing for vias or low-aspect-ratio vias are necessary. Conventional processing therefore places restrictions on circuits packing density. The difficulties met using these traditional processes lead to their replacement with high-aspect-ratio, anisotropically etched vias and contact holes and use of chemical vapor deposited (CVD) tungsten (W) to form plugs [2].

Following the CVD W deposition, a blanket W removal process is required. A plasma etchback process has been used for blanket W removal in the early process flow due to availability of the equipment and process chemistry [2 – 5]. However, the W plasma etchback process has inherited several problems due to its design: (1) high defect density due to the gas used in the plasma, such as SF_6, which is a highly polymer generating gas. (2) Due to the poor selectivity to oxide by using SF_6 plasma chemistry, the process requires high selectivity of W to liner metal – usually TiN. (3) Additional overetch is required to compensate for etch non-uniformity. This results in severe W recess in the plug area. (4) severe loading effect on plug recess, due to pattern density.

Because the problems associated with W plasma etchback process are inherited in its chemistry and process design, it was difficult to resolve them [2-5]. Therefore, an alternative process – metal chemical mechanical polishing – has been proposed and studied extensively in the last several years [6-12]. With the maturing equipment and consumable sets, the W CMP process is proven to be better in terms of defectivity [6], controlling the plug recess, and high selectivity to oxide over traditional W plasma etchback process. However, W CMP is still relatively a new process. It has faced several challenges in process integration, equipment and consumable stability, and defectivity. In this paper, a new W CMP process targeted in those issues is described in details. The interactions between pre-tungsten CMP process on W polish and the improvement in across-wafer uniformity during W CMP process were discussed. A head-to-head comparison of the optimized KIO_3 process with the standard $Fe(NO_3)_3$ process in cost of ownership of the equipment, defectivity, and in-line process control and monitor parameters is described. Finally, the significant positive effect of the optimized KIO3 tungsten CMP process on interconnect integration and die sort yield are reported.

EXPERIMENT

There were two types of wafers used in this experiment: blanket test wafers and fully integrated production wafers. The blanket test wafers used in this experiment were P(100) silicon deposited with 7500Å of TEOS oxide followed by a sputtered TiN adhesion layer and a deposited 5000Å W film. The tungsten film used in this experiment was deposited in an Applied Materials CENTURA reactor, using WF_6 reduced with SiH_4 and H_2. After W deposition, W CMP was performed using different hardware and process parameters.

Fully integrated wafers used in this study were processed with typical sub-0.35µm CMOS process flow. The first interconnect module for this experiment started with a 3-layer oxide deposition of the first inter-metal-oxide (IMO). A layer of 1500Å undoped TEOS was used as base oxide, on top of it was 3200Å BPTEOS (densified after deposition), and it was capped with 8000Å Si-rich Silane-based oxide. It was followed by a dielectric CMP process which was targeted for 3000Å oxide removal. After contact mask, etch and strip processes, the contact metalization consisted of an adhesion layer with 700Å Ti and 500Å TiN, followed by 5000Å CVD tungsten deposition. The W deposition was processed with the same equipment and process parameters as the blanket wafers mentioned above. Subsequently, after the W CMP process, the first metal layer was deposited, masked, and etched. This completing the first IMO module. The second and third interconnect modules have similar process flows, with some films stack changes. The IMO stack was changed to TEOS, SOG and TEOS. The adhesion layer before W deposition was using TiN only, therefore, the W CMP process changed accordingly.

The polisher used in this study was a Strasbaugh 6DS-SP dual-spindle, dual-platen system. It was equipped with high-throughput tray, inferred temperature sensor, primary platen temperature control, and external endpoint computer. The endpoint computer has the capability of monitoring, recording and controlling the polishing process using signals from head motor currents, table motor current and pad temperature. The system also has three separated slurry dispense lines, which allow to dispense the metal slurry on the primary pad and oxide slurry on the secondary pad used in oxide buff process. Two types of wafer carriers were studied extensively in this experiment. Standard wafer carrier has a retaining ring used for retaining the wafer and a clamp ring used for keeping the retaining ring in place. There was a special hole pattern in the wafer carrier which can be either blocked and open. As a result, the back pressure of a wafer can be controlled independently from the down force (Fig. 1). The fixed-ring ViPRR (viable pressure retaining ring) wafer carrier has only one fixed retaining ring for keeping the wafer in place. It has similar hole pattern design as the that for the standard carrier, which can also be either blocked or opened to control the back pressure independent from the down force (Fig. 2).

The two tungsten slurries evaluated in this study. The first slurry Cabot W2000 has $Fe(NO_3)_3$ as oxidizer and Al_2O_3 as abrasive. The second slurry Rodel MSW2000 has KIO_3 as oxidizer and Al_2O_3 as abrasive. Each slurry has two components, which were mixed based on manufacturing recommendations. The polishing pads used in the experiments were TWI 817 for W2000 slurry and Rodel IC1000/Suba4 Stack for MSW2000.

The metal films polish rates (Ti, TiN, and W) were measured on Prometrix 4-point resistivity probe. Oxide loss on blanket and fully integrated wafers were measured on KLA-Tencor UV1280. In-line SEM was used for defect inspection.

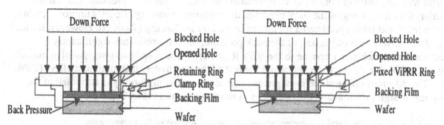

Fig. 1. Schematic diagram for the standard wafer carrier used in Strasbaugh 6DS-SP for this study.

Fig. 2. Schematic diagram for the fixed-ring ViPRR (Viable Pressure Retaining Ring) wafer carrier used in Strasbaugh 6DS-SP for this study. Noticed the single piece retaining ring used in this type wafer carrier.

RESULTS AND DISCUSSION

The study was intended to look at the effect of process and hardware parameters on the W polish rate and across-wafer uniformity. The correlation between the across-wafer uniformity of the tungsten polish process and the die sort yield was examined. In order to understand the polish rates at the extreme wafer edge, the film thickness was measured across the wafers. The same technique was used for both blanket and fully integrated wafers. As a result, the data from blanket wafers can be correlated with the results from fully integrated wafers.

Polish rate and across-wafer uniformity data

Unlike W plasma etch back process, the typical W CMP process usually removes the adhesion layer such as Ti/TiN or TiN during the primary polish. As a result, during the over polish step there is some oxide loss. Since the oxide deposition, planarization CMP (oxide CMP), and tungsten CMP steps are subsequent to each other, the oxide thickness profile could become worse further into the process flow. Therefore, the across-wafer non-uniformity of the oxide loss during W CMP process is one of the very important process parameters needs to be optimized. To determine the effect of the process and hardware parameters on the polish rate and the across-wafer uniformity, designed experiments were run and trends were determined using analysis of variance techniques. Table speed, wafer carrier speed, down force, back pressure, blocked hole pattern, and carrier types were examined for their effects on polish rate and across-wafer uniformity. The variable ranges encompassed by the experiments used in this study are summarized in Table I.

The polish rates obtained were in the range of 2000 – 8500 Å/min. The across-wafer uniformity was calculated in percentage standard deviation of polish rate, ranged from 1.7% to 23.4%. In summary, the polish rate trends follow the theoretical model using Preston's equation [13]: the higher the down force and the higher the table speed, the faster the polish rate. It was true for both wafer carrier types. However, the across-wafer uniformity is significantly different for different wafer carrier types. For standard wafer carrier, the across-wafer uniformity was ranging from 4.7% to 23.4%. On the other hand, the across-wafer uniformity for ViPRR carrier was ranging from 1.7% to 9.3%, which is significantly better than that of the standard carrier. There was also another noticeable difference in the effect of the back pressure. For ViPRR carrier, with increasing of the back pressure, the uniformity was significantly improved, while down force and table speed had a minimum impact on uniformity (Table II). However, for the standard carrier, with increasing of the back pressure, the uniformity has significantly deteriorated, which was opposite to the ViPRR carrier. Table speed had some positive impact on uniformity for standard carrier, while the rest of the parameters had minimum effect (Table III). The different effect of the back pressure on across-wafer uniformity is attributed to the physical design difference in the two types of the wafer carriers (Fig. 1 and 2).

Table I. List of experimental parameters varied and ranges studied in the experiments.

Variable	Range
Table Speed (rpm)	65 - 105
Carrier Speed (rpm)	70 - 113
Down Force (psi)	4 - 7
Back Pressure (psi)	0 - 5.1
Carrier Type	STD - Fixed Ring ViPRR
Carrier Blocked-Hold Pattern	STD - No-Hold blocked

Table II. General process trends for W CMP for ViPRR type wafer carrier. Up arrows mean that the parameter increases as a function of an increase in the process variable.

Variable	W Polish rate	Uniformity
Table Speed (rpm)	↑	↓
Carrier Speed (rpm)	↔	n/a
Down Force (psi)	↑	↓
Back Pressure (psi)	↔	↓
Carrier Blocked-Hold Pattern	↔	↔

Table III. General process trends for W CMP for standard type wafer carrier. Up arrows mean that the parameter increases as a function of an increase in the process variable.

Variable	W Polish rate	Uniformity
Table Speed (rpm)	↑	↓
Carrier Speed (rpm)	↔	↔
Down Force (psi)	↑	↔
Back Pressure (psi)	↓	↑
Carrier Blocked-Hold Pattern	↔	↔

Slurry Comparison

W2000 W CMP slurry has $Fe(NO_3)_3$ as oxidizer which is more chemically reactive. As a result, high selectivity to oxide can be achieved with this slurry. However, because it is chemically reactive, the W plug recess is more severe. And also due to its electrochemical property, the slurry residue tends to accumulate around recessed area, such as recessed W plugs, alignment marks (Fig. 3). A buff process using oxide slurry is usually used to remove certain amount of oxide in order to minimize the amount of plug recess and remove the slurry residue left on the wafer surface. The additional oxide buff process required for W2000 slurry results in higher oxide loss, deteriorating the across-wafer uniformity, and slowing down the equipment throughput. Higher oxide loss and additional across-wafer non-uniformity introduced in W CMP can significantly affect the overall oxide thickness profile across the wafer as further into process flow [14]. Moreover, because Fe(NO$_3$)$_3$ is highly aromatic, it results in a film being deposited on the equipment. Due to low pH, the slurry is also very aggressive on the tools. As a result, the consumable lifetime is limited in order to maintain product defectivity. Consequently, a very labor-intensive preventive maintenance is required [15].

On the other hand, the W CMP slurry that uses KIO$_3$ as oxidizer such as MSW2000, is less chemically reactive comparing with W2000. As a result, the selectivity to oxide for MSW2000 is not as superior as that of W2000. However, due to it electrochemical property, the KIO$_3$ based slurry tends to have clean wafer surface even without oxide buff process after the primary polish (Fig. 4). By improving the across wafer uniformity, eliminating the oxide buff process and using automated endpoint system, the overall oxide loss during W CMP process can be precisely controlled.

Automated Endpoint Algorithm

Since MSW2000 uses KIO$_3$ as oxidizer, it tends to be more mechanical in nature than W2000, which is more chemically reactive. As a result, it enables the capability to endpoint the polish process by monitor the primary table motor current for the KIO$_3$ slurry. Fig. 5 and 6 have shown the examples of the primary table current during the W CMP process of polishing Ti/TiN/W and TiN/W wafers using the KIO$_3$ slurry. The extra peak noticed in Fig. 5 represent the clearing of the Ti layer due to more mechanical removal nature of this film.

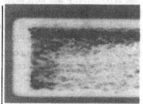

Fig. 3. Heavy slurry residue accumulated in the recess area (alignment mark) after standard Fe(NO$_3$)$_3$ W CMP process.

Fig. 4. No slurry residue found in the recess area (alignment mark) after the KIO3 W CMP process with the new ViPRR carrier.

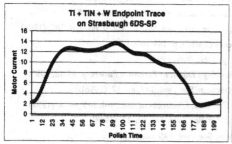

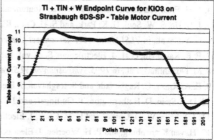

Fig. 5. Using primary table motor current as the process control for endpointing the W primary polish process. Example of the endpoint trace for Ti/TiN/W stack wafers.

Fig. 6. Using primary table motor current as the process control for endpointing the W primary polish process. Example of the endpoint trace for TiN/W stack wafers.

Die Sort Yield Result

With the improvement in across-wafer uniformity using the ViPRR carrier and automated endpoint algorithm for KIO_3 slurry (MSW2000) on IC1000 pad, the die sort yield was significantly higher compared to the die sort yield for the $Fe(NO_3)_3$ (W2000) slurry with standard carrier process (Table IV). Notice the significant sort yield difference from the edge of the wafers due to the improvement of the across-wafer uniformity during the W CMP process. There was no die yield difference between the two over polish splits, which verified the robustness of the automated endpoint algorithm.

Table IV. Normalized die sort yield comparison for standard $Fe(NO_3)_3$ (W2000) slurry with standard wafer carriers on Strasbaugh and new KIO3 (MSW2000) slurry with the ViPRR type wafer carriers.

Processes	Normalized Die Sort Yield	Normalized Deep Edge Yield
Fe(NO3)3 + Standard Carrier	21.8	45.6
KIO3 + ViPRR Carrier at Endpoint	38.2	56.0
KIO3 + ViPRR Carrier + 25% Over Polish	40.2	56.8

CONCLUSION

A robust, production worthy W CMP process using the KIO_3 slurry and new style (ViPRR) wafer carrier was developed. With the KIO_3 electrochemical property, the oxide buff process was eliminated. By using a reliable automated endpoint algorithm, the overall oxide loss during W CMP process is minimized. With the help of the new design of the ViPRR wafer carriers, the best across-wafer uniformity process was developed for the W CMP process. As a result, a significant improvement in the sort yield has been observed for the new process.

REFERENCE

1. S. Bothra, et al., Solid State Technology, February, 1997, p. 77.
2. L. R. Allen and J. M. Grant, J. Vac. Sci. Technol. B 13(3), p. 918, May/Jun 1995.
3. C. A. Bollinger, et al. in *VLSI Multilevel Interconnect Conference (VMIC)*, p. 21-27, 1990.
4. J. M. F. G. van Laarhoven, H. J. W. van Houtum and L. de Bruin in in *VMIC*, p. 129-135, 1989.
5. J. H. Ha, S. W. Seol, H. K. Park, and S. H. Choi, IEEE Trans. on Semi. Manuf., vol. 9, no. 2, p. 289, (1996)
6. C. Yu, et al., in *VLSI Multilevel Interconnect Conference (VMIC)*, p. 144-150, 1994.
7. F. B. Kaugman, et al., J. Electrochem. Soc., vol. 138, p. 3460, Nov. 1991.
8. M. Norishima, et al., in *1995 Symposium on VLSI tehcnology Digest of Technical Papers*, p. 47-48, 1995.
9. E. A. Kneer, C. Raghunath, and S. Raghavan, J. Electrochem. Soc., vol. 143, no. 12, (1996).
10. D. J. Stein, D. Hetherington, T. Guilinger, and J. L. Cecchi, J. Electrochem. Soc., vol. 145, no. 9, (1998).
11. J. J. Shen, W. D. Costas, and L. M. Cook, J. Electrochem. Soc., vol. 145, no. 12, (1998).
12. H. van Kranenburg, and P. H. Woerlee, J. Electrochem. Soc., vol. 145, no. 2, (1998).
13. F. Preston, J. Soc. Glass Tech., 11, 214, (1927).
14. A. H. Liu, R. Solis, and J. H. Givens in *Technical Symposium in CMP Process SEMICON China*, 1999.
15. W. Parmantie, J. H. Givens, in *1ˢᵗ Intl. Conf. On Adv. Materials and Processes for Microelectronics.*, 1999.

XPS and Electrochemical Studies on Tungsten-Oxidizer Interaction in Chemical Mechanical Polishing

Dnyanesh Tamboli, Sudipta Seal and Vimal Desai
Advanced Materials Processing & Analysis Center (AMPAC)
Mechanical, Materials & Aerospace Engineering Department
University of Central Florida
Orlando, FL 32816

ABSTRACT

Electrochemical interaction between the oxidizer and the metal is believed to play a key role in material removal in tungsten CMP. In this study, we use X-ray Photoelectron Spectroscopy (XPS) in conjunction with electrochemical measurements in both in-situ polishing conditions as well as in static solutions, to identify the passivation and dissolution modes of tungsten. Dissolution of tungsten oxides was found to be the primary non-mechanical tungsten removal mechanism in CMP.

INTRODUCTION

Chemical mechanical polishing (CMP) is the enabling technology for sub 0.35 micron size multilevel semiconductor device fabrication. Tungsten CMP is used extensively in the process to form contacts and vias with damascene architecture. Tungsten CMP mechanism proposed by Kauffman et al.,[1] describes tungsten removal to be result of continuous cycles of (a) blanket passivation (WO_3 oxide growth) on entire surface (b) abrasion of passive oxide films from high regions on the surface (c) enhanced ionic dissolution from the high regions due to loss of passive film (d) dynamic repassivation of the abrading surface. According to this mechanism, ionic dissolution from the high regions undergoing abrasion is believed to be the primary mode of tungsten removal. This study used a slurry (pH 6) containing $K_3Fe(CN)_6$ as an oxidizer. Some studies have supported this mechanism by showing that corrosion current measured during polishing is proportional to actual polishing rates.[2,3] In contrast to these reports, there has been some serious criticism to this mechanism from Stein et al.,[4] and Kneer et al.[5] Stein et al.,[4] have compared the electrochemical dissolution currents during polishing with actual removal rates in same sample. They observed that tungsten dissolution rates as measured by electrochemical techniques are about 1 to 2 orders of magnitude lower than the actual polishing rates. Also most interestingly, the polishing rates under cathodic polarization conditions were not significantly lower than those obtained under normal open circuit potential (OCP) conditions. Based on these observations it was concluded that passive oxide formation is not an important step in tungsten removal in tungsten CMP. However, no alternate material removal mechanism was proposed in this study. Kneer et al. have also reported that the dissolution currents are too low to account for the total tungsten removed. Based on AFM studies of polished tungsten wafers, they have suggested a possibility of corrosion assisted fracture to be a dominant material removal mode.

Whether a passivating film forms on the surface during polishing is one of the key issues in understanding the tungsten CMP mechanism. However neither supporters or the critics of the mechanism have supplied any surface analysis data to support their claims. Our study aims at providing a new light into the issues of passive film formation and ionic dissolution of tungsten in CMP using surface sensitive X-ray Photoelectron Spectroscopy (XPS) and a variety of electrochemical techniques. Our approach is to apply the recent advances in the theories of passivation to understand passivity and dissolution modes in tungsten in both static and CMP conditions.

EXPERIMENTAL

XPS studies were performed on 1.5cm X 1.5cm pieces cut from a 8" diameter silicon wafer coated with 4000 A^0 thick LPCVD tungsten layer. Most of the electrochemical studies were performed on a 1" diameter bulk tungsten electrode (99.95% purity, purchased from Target Materials). This electrode was mirror polished for use in electrochemical studies. Using XPS it was observed that as received tungsten coated wafers had an approximately 15 to 20 A^0 thick native oxide layer present on the surface. Sample preparation, therefore involved immersing the samples in 45% KOH solution for 5 minutes to remove these native tungsten oxides from the surface. All the chemicals used in this study were of reagent grade purchased from Fisher Scientific. The pH of the solutions used in these experiments was maintained using various buffer solutions purchased from Fisher Scientific.

We used both in-situ and ex-situ electrochemical measurements to understand tungsten-oxidizer interaction. Ex-situ electrochemical studies were performed in model EG&G Electrochemical Flat Cell using EG&G potentiostat model 273. CMP of tungsten was performed using a Buehler Minimet 1000 system. In this set-up, polishing pad is stationary and the polishing piece is moved relative to pad causing the abrasion. Polishing experiments were carried out at 4 psi pressure and 30 rpm speed (equivalent to approximately 20 cm/sec linear velocity). The polisher was modified to incorporate in-situ electrochemical measurements. A 1" diameter bulk tungsten sample was used as the working electrode, whereas the stainless steel polishing platen was used as the counter-electrode. Ag/AgCl reference electrode with a vycor tip was used as a reference electrode. The slurries used in these experiments were prepared by diluting the abrasive component of RODEL MSW2000 slurry to 4.3 wt % using buffer solutions containing oxidizers with desired concentrations.

XPS was performed using a PHI 5400 ESCA system. Samples were transferred in an inert argon atmosphere to the XPS analysis chamber to minimize the oxide formation on surface due to exposure to air. Non-monochromatic Al Kα X-ray source (hν= 1486.6 eV) at a power of 350 watts was used for the analysis. The spectrometer was calibrated using Au 4f 7/2 peak at 84±0.1 eV (FWHM of 0.8 eV). Standard data fitting procedures were observed for fitting various core level peaks with component peaks having a Gaussian-Lorentzian distribution. Binding energy reference has been made with respect to C (1s) peak at 284.6 eV.[6]

RESULTS AND DISCUSSION

Macdonald et al.[7] have developed an elaborate theory for formation of oxide films on various metals based on movement of point defects (ionic vacancies) through the lattice. There are five different processes that need to take place during steady state oxide formation (1) Generation of cation vacancies at the oxide/solution interface because of injection of cations into the solution. These vacancies move towards metal/oxide interface under electric field. (2) Formation of oxygen vacancies at metal/oxide interface through a non-conservative lattice process. The vacancies move under electric field towards oxide/solution interface. (3) Injection of a metal atom into a vacant cation site at the metal/oxide interface. (4) Annihilation of oxygen vacancies at the oxide/solution interface (5) Dissolution of the oxide at oxide/solution interface. This process also does not conserve lattice. At the steady state, the thickness of the oxide film does not change with time, thus indicating that the processes 2 and 5 balance each other at the steady state. The growth mechanism of the oxide is dictated by the relative flux rates of anion and cation vacancies. These rates are in turn governed by the ease of formation of vacancies, which are dependent on the valency of the cations. In WO_3, cations exist in hexavalent state. Generation and movement of an hexavalent vacancy is much more difficult that that of a bi-valent anion vacancy. Therefore, WO_3 film growth is dominated by oxygen vacancy movement. In other words, WO_3 is an anion conducting film. According to this point defect model for oxide growth, for anion conducting films the current density in a oxide is independent of the applied potential as long as the valency of the metal does not change while going from oxide to dissolved ionic species. In these oxides steady state current is determined by the balance between the flux of oxygen vacancies and the dissolution of oxides at the oxide/metal interface. Since dissolution of oxides is *pH* controlled, anion-conducting oxides show a strong dependence of steady state current density on *pH*.

In this study we applied these diagnostic criterion to evaluate whether the oxide films on tungsten are anion conducting in oxidizer solutions. As expected from the potentiodynamic polarization experiments, which show a constant current passivation regime, the steady state current density was found to be independent of potential in passive region using potentiostatic measurements.

Figure 1 shows the effect of pH on the steady state current density. It can be seen that the current density is minimum in the pH range of 2-3. The current increases exponentially with the pH changes in either direction. The reactions at lower pH are believed to be;

$$WO_3 + 2H^+ \rightarrow WO_2^{2+} + H_2O \qquad (1)$$

Whereas at higher pH the reaction is likely to be dominated by

$$WO_3 + 2H^+ \rightarrow WO_2^{2+} + H_2O \qquad (2)$$

The current densities in KIO_3 are lower than in $K_3Fe(CN)_6$ solutions. The pH of minimum potential is also shifted to somewhat higher pH in KIO_3 solution. The difference in their behavior may be attributed to ionic radii of IO_3^- and $Fe(CN)_6^{3-}$ ions. IO_3^- ions have much larger size and are thus may provide a hindrance to the anodic dissolution reactions. The value of $\delta ln(I_{ss})/\delta pH$ is about 1.45. This is lower than the value of 2.303 obtained by Macdonald et al.[7] This study was carried out in static environment, unlike the rotating disk electrode set-up used by MacDonald et al. The differences in the value of this slope may probably be attributed to the differences in mass transfer properties in solution.

XPS Analysis

In an earlier study we had reported the XPS analysis of tungsten oxides formed during anodic polarization experiments[8]. It was determined that even at high applied potentials, the oxide thickness values are less than the mean free path of electrons in the oxides (generally assumed to be between 30 to 50 A^0). Clearly the oxide growth in tungsten is a slow process. However, despite the relatively small thickness values, the steady state current density during anodic polarization is restricted to a few tens of microamperes.

Figure 2 compares the tungsten 4f spectra of tungsten exposed to different solutions for 10 minutes in open circuit potentials in static environment. The etched condition refers to the standard treatment given to all samples to remove native oxide layer from the surface. There is very little difference in W (4f) spectra in the four samples investigated, inspite of very different electrochemical behavior observed in these solutions. For example, the OCP potential of tungsten in 2% $K_3Fe(CN)_6$ solution at pH 4 lies in the passive regime as observed by D.C.

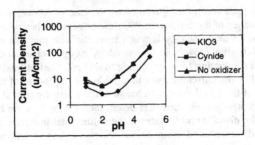

Figure 1: Effect of pH on steady state current density during anodic polarization at 3 V.

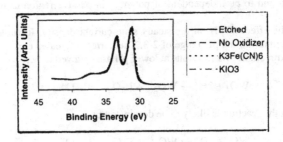

Figure 2: W(4f) XPS spectra taken on various tungsten samples immersed in various solutions for 10 minutes.

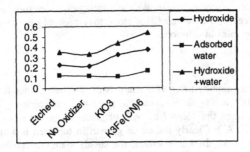

Figure 3: Contributions to O(1s) peak in various components in tungsten samples immersed in various solutions for 10 minutes.

polarization measurements. Tungsten in KIO$_3$ solution has much lower corrosion currents than in K$_3$Fe(CN)$_6$ solutions, indicating that KIO$_3$ is a less aggressive oxidizer. As expected the buffer solution showed insignificant corrosion currents. Although no difference in W spectra are observed a substantial difference was observed in O 1s spectra. O 1s peak was deconvoluted into three peaks, oxide peak in the range of 530.3-530.6 eV, chemisorbed –OH peak at 532 eV and adsorbed water peak in 533-533.5 eV range. As seen in figure 3, hydroxyl groups and adsorbed water peaks increased with the oxidizer addition, with the increase in maximum for K$_3$Fe(CN)$_6$, which was found to be the more potent oxidizer of the three used in this study, thus highlighting the role of hydroxyl ions in formation and dissolution of oxide films.

To evaluate the effect of polishing on surface chemistry of tungsten coated wafers, LPCVD tungsten were polished under different conditions, (a) using both with both slurry and oxidizer (b) pad polish with oxidizer solution only (c) polishing using slurry without an oxidizer. W(4f) spectra of samples polished on Suba4/IC1000 stacked pad in the absence of abrasives are shown here with the in figure 4. Even though curve fitting and deconvolution detected WO$_3$ and WO$_{2-x}$ type of oxides, the oxide thickness was hardly different from that observed in wafers undergone etching to remove native oxides. This indicates that even though oxide growth requires considerable driving force, oxide monolayer formation is probably an instantaneous process.

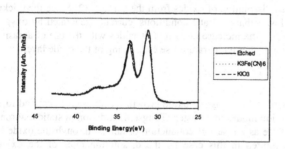

Figure 4: W(4f) XPS spectra for tungsten polished without abrasive particles in the two oxidizer solutions compared with the spectrum on the sample given etch treatment to remove oxides.

In-situ Electrochemical Polarization Studies

To isolate effects of oxidizers and abrasives, electrochemical polarization experiments were performed using following sets of conditions (a) polishing without any abrasive or oxidizer (b) polishing with oxidizer solution but no abrasive (c) polishing using abrasives, but without oxidizer (d) polishing using both abrasives and oxidizers. These studies were carried out at pH 4 using 2% K$_3$Fe(CN)$_6$ as the oxidizer. The potentiodynamic polarization scans taken at the scan rates of 5 mV/s are shown in figure 5. It can be seen that although the use of abrasive increases the pseudo steady state current density in the passivation regime, this increase is only marginal.

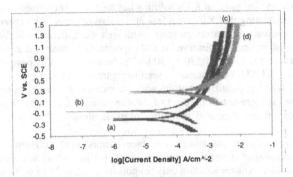

Figure 5: Insitu electrochemical potentiodynamic scans on tungsten polished under various conditions (a) pH 4 buffer solution containing abrasive particles, (b) pH 4 solution without containing abrasives, (c) pH 4 $K_3Fe(CN)_6$ solution without abrasive, (d) pH 4 $K_3Fe(CN)_6$ solution containing abrasive particles

This is consistent with the observation that there is virtually no growth of oxides during just pad abrasion. The increased current densities in passivation regime can be explained in following way (1) Localized temperature increases and drastic reduction in diffuse layer width during polishing leads to enhanced dissolution of oxides from the surface. (2) Since the thickness of the oxides involved is much lower than in static conditions, currents generated by oxygen vacancy flux is likely to be very high. The increase in current densities with the use of abrasive particles may be attributed to either a further temperature rise or thinning of the oxide layer.

CONCLUSIONS:

In this study, point defect model for passive oxide growth on tungsten is applied to understand the passivation and dissolution modes in tungsten. Tungsten oxidation in static environments was determined to be taking place as a result of conduction of anions through the oxide lattice. The steady state currents are limited in this case by the dissolution rates of the oxides. During polishing in the absence of abrasives, the current densities were 1 to 2 orders magnitude higher than the static conditions. Presence of abrasives only slightly increased the dissolution currents. Since it is unlikely that mechanical abrasion of metal takes place in the absence of abrasives, enhanced tungsten oxide dissolution and not the ionic dissolution from abraded surface may be the primary non-mechanical material removal mechanism in tungsten CMP.

ACKNOWLEDGEMENTS:
Authors would like to thank Lucent Technologies, Orlando FL and I4 High Tech Corridor Initiative for financial support of this project. We would also like to thank Dr. A. Maury for his inputs.

REFERENCES:

1. F. B. Kaufman, D. B. Thompson, R. E. Broadie, M. A. Jaso, W. L. Gutherie, D. J. Pearson and M. B. Small, *J. Electrochem. Soc.*, **138**, 3460 (1991)
2. C. C. Streinz, D. Ligocki, T. Myres, V. Brusic, in "Chemical Mechanical Planarization", eds. I. Ali and S. Raghavan, *The Electrochemical Society Proceedings* Vol. **96-22**, Publ. The Electrochemical Society, NJ, 159 (1996).
3. J. Farkas, R. Caprio, R. Bajaj, C. Galanakis, R. Jairath, B. Jones and S. Tzeng, in, "Advanced Metallization for ULSI Applications in 1994", eds. R. Blumenthal and G. Janssen, publ. *Materials Research Society*, PA, 25 (1995).
4. D. Stein, D. Hetherington, T. Guilinger and J. Cecchi, *J. Electrochem. Soc.*, **145**, 3190 (1998)
5. E. A. Kneer, C. Raghunath, V. Mathew, S. Raghavan and J. S. Jeon, *J. of Electrochem. Soc.*, **144**, 3041 (1997).
6. T. Barr, S. Seal, *J. Vac. Sci. Tech.*, **A13**, 1239 (1995).
7. D. Macdonald, S. Biaggio and H. Song, *J. Electrochem. Soc.*, **139**, 170(1992)
8. D. Tamboli, S.Seal, V.Desai and A. Maury, to be published , J. Vac. Sci. Tech. (accepted 1999)

TRIBOLOGICAL EXPERIMENTS APPLIED TO TUNGSTEN CHEMICAL MECHANICAL POLISHING

MARC BIELMANN*, UDAY MAHAJAN*, RAJIV K. SINGH*, PANKAJ AGARWAL**, STEFANO MISCHLER**, ERIC ROSSET**, AND DIETER LANDOLT**

*Department of Materials Science and Engineering, University of Florida, Gainesville, FL 32611
**Department of Materials Science, Ecole Polytechnique Federale Lausanne, Lausanne, Switzerland

ABSTRACT

Tungsten CMP involves a synergistic interaction of electrochemical and tribological (wear) phenomena. So far, numerous studies have been conducted using static electrochemical measurements as well as some polishing experiments. In this study, we present some results obtained from carrying out potentiodynamic measurements and tribological experiments in a reciprocating sphere-on-plate tribometer, which allowed a precise control of mechanical and electrochemical conditions. In addition, anodic current-time transient measurements were also used to characterize the kinetics of tungsten passivation reaction. These results indicate that the presence of an passive film is essential for wear of tungsten to take place.

INTRODUCTION

Chemical Mechanical Polishing (CMP) still remains a poorly understood process, although it has been integrated quite successfully into the industry. CMP of metals like tungsten involves strong interactions between electrochemical phenomena and mechanical abrasion. The most widely accepted model for tungsten CMP was developed by Kaufman et al. [1], which postulates that material removal takes place by the formation of a passive oxide film, which is then abraded by the particles in the polishing slurry. Several electrochemical studies have also been performed on the tungsten system [2],[3], along with measurements carried out during CMP [4], [5]. Many of these studies report contradictory results, with Kneer et al. [4] claiming that removal during CMP is the result of corrosion assisted fracture, and Stein et al. [5] concluded that a passive film on the tungsten surface is not required for CMP. These different conclusions have left a lot of unresolved questions about the nature of the removal process. To shed some additional light on the tribochemical phenomena occurring during tungsten CMP, we conducted electrochemical measurements and wear experiments using a sphere-on-plate tribometer system.

EXPERIMENT

The sliding wear experiments were conducted by rubbing an alumina sphere (6 mm diameter) against a tungsten disk (1.3 cm diameter and 0.25 cm thickness). The experiments were carried out at the EPFL (Ecole Polytechnique Fédérale de Lausanne or Swiss Federal Institute of Technology) in the laboratory LMCH (Laboratoire de Métallurgie Chimique). The samples were cut from a tungsten rod of 1.3cm diameter and 99.95% purity procured from Alfa Aesar®. The disks were embedded in a plastic resin (Technovit®) with an electrical contact allowing the connection. Then the disks were mirror polished using 1 μm diamond paste.

Potentiodynamic polarization curves were measured to characterize the electrochemical behavior of tungsten in 0.5 M H_2SO_4. A tungsten stationary disk electrode of 1.27 cm diameter (1.266 cm^2 surface area) was used in the setup described below (Fig. 1). The reference electrode was Ag/AgCl and the scan rate was 5 mV/s. In addition, anodic current-time transients were recorded to measure passive film formation and growth. These experiments were carried out in the same cell as the potentiodynamic polarization measurements using a HEKA potentiostat (rise time < 5 μs) and a Macintosh-computer-based data acquisition system recording current and potential values with a sampling rate of 10 kHz with 16-bit resolution. The procedure involved a cathodic polarization for 1 min at a potential of – 2 V versus Ag/AgCl reference electrode to remove surface oxides. Then the potential was increased by a step to the chosen value and the current was recorded.

Frictional tests were carried out using a reciprocating sphere-on-plate tribometer permitting the control of mechanical and electrochemical conditions. Sliding wear conditions are established by rubbing an alumina sphere fixed on a vertical shaft against a fixed flat plate sample immersed in the test solution. The electrode potential of the plate is controlled using a three-electrode setup in conjunction with a potentiostat, including a platinum wire counter electrode and a Ag/AgCl reference electrode. The electrochemical cell is placed on a load cell, which measures the applied normal force as shown in Fig. 1. The frictional force is measured by means of a piezoelectric force transducer. Reciprocal sphere motion is provided by a linear motor driven by a triangular wave form signal generator.

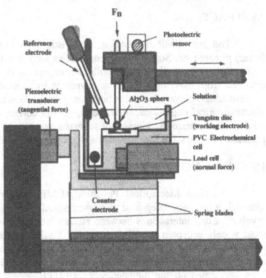

Fig. 1: Schematic representation of experimental arrangement for tribocorrosion experiments with alumina sphere, load cells and electrochemical cells

Both the vertical and horizontal sphere displacements can be measured using a photoelectric sensor. During the wear experiments the frictional and the normal forces, the vertical (linear wear) and horizontal sphere displacement, as well as the electrochemical parameters (current and potential) is continuously monitored using a Macintosh computer running on LabView-Language-based software developed in LMCH of the EPFL.

The sphere was oscillated at a frequency of 5 Hz. The linear motor maintained the pin motionless for 20 μs at the end of each stroke. In this way the stroke length of 5 mm corresponded to a sliding speed of 62 mm/s. The applied load was 5 N, resulting in a starting Hertzian pressure of ~ 1.7 MPa.

The following types of electrochemical conditions were applied:
    (i)   Free potential (OCP) in 0.1 M $K_3Fe(CN)_6$ at pH 3,
    (ii)   Free potential in 0.5 M H_2SO_4,
    (iii) Applied cathodic potential (- 0.9 V versus Ag/AgCl) in 0.5 M H_2SO_4,

(iv)   Applied anodic potentials: -0.4 V, -0.2 V, 0 V, + 0.2 V, + 1.2 V, + 1.7 V versus
        Ag / AgCl reference electrode.

For these experiments, a cathodic pretreatment was made first by polarizing at – 2 V versus
Ag/AgCl for 1 minute, followed by polarizing at + 2 V versus Ag/AgCl for 3 minutes. And then
the preselected potential for the experiments was applied and the rubbing was started. The
pretreatment procedure ensured that each experiment started with a similar surface.

All experiments were carried out at room temperature (21-22°C) and the rubbing time
was 25 minutes, corresponding to 15,000 strokes. At the end of the test, the tungsten plates and
the alumina spheres were removed from the solution and rinsed with distilled water. For each
condition the experiments were repeated twice or three times. The wear scar volume was
determined by optical profilometry using a UBM laser system.

RESULTS AND DISCUSSION

Fig. 2 shows a typical polarization curve and one can identify the active, passive and
transpassive potential region. In the active region, several peaks can be observed which can be
attributed to different oxidation states.

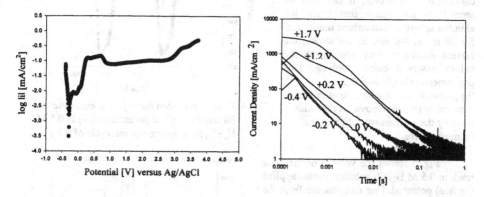

Fig. 2: Potentiodynamic polarization
        curve of Tungsten in 0.5 M H_2SO_4

Fig. 3: Potentiodynamic current transients
        measured on cathodically prepolarized
        tungsten electrode in 0.5 M H_2SO_4 at
        several potentials indicated on the plot
        (vs. Ag/AgCl)

Fig. 3 shows the potentiodynamic current transients as a function of time on a logarithmic
scale for the following applied potentials versus Ag / AgCl reference electrode: -0.4 V, -0.2 V, 0
V, +0.2 V, + 1.2 V, + 1.7 V. One can see from the figure that the current decreases gradually due
to passive film formation and growth. This figure also shows that the charge used for passivation
is increasing with increasing anodic applied potential. In order to relate the anodic charge
densities measured in these experiments and the charge required to passivate a bare metal
surface, it should be verified to what extent the cathodic treatment effectively eliminates the

99

native oxide film covering the metal surface. Thus, we will consider these measured anodic charges qualitatively, but not quantitatively. However, if we compare the charge by integrating the current transients up to 0.1 s. and 0.01 s., which correspond to an abrasion frequency of 10 Hz and 100 Hz respectively, the anodic charges have the same order of magnitude values. The abrasion will take place 10 times more often in case of 100 Hz frequency compared to 10 Hz. Therefore if we assume that wear happens by oxide film removal as described by Kaufman's model [Kau91], an abrasion frequency of 100 Hz is expected to remove more material than abrasion under 10 Hz frequency. Moreover, the applied potential is changing the oxide thickness, so removal rate is expected to increase with anodic applied potential and to be negligible on applying cathodic potential if we assume that removal is based on Kaufman's model. In order to check these hypotheses, several wear experiments have been conducted and results will be presented in the next section.

Fig. 4 shows the current recorded during a sphere motion cycle of 0.2 s (back and forward motion) in 0.5 M H_2SO_4 with an applied potential of 2 V versus Ag / AgCl. Before starting the experiment, the current was in the range of 0.04 mA. Thus one can observe that during rubbing, the current is increasing considerably. However, it can also be seen from the figure that when the alumina sphere is maintained motionless for 20 ms at the end of the stroke, the current decreases very rapidly to the values observed before rubbing. This phenomenon clearly indicates the passivation of the surface after the removal or partial removal of the oxide layer by the mechanical action of the alumina sphere.

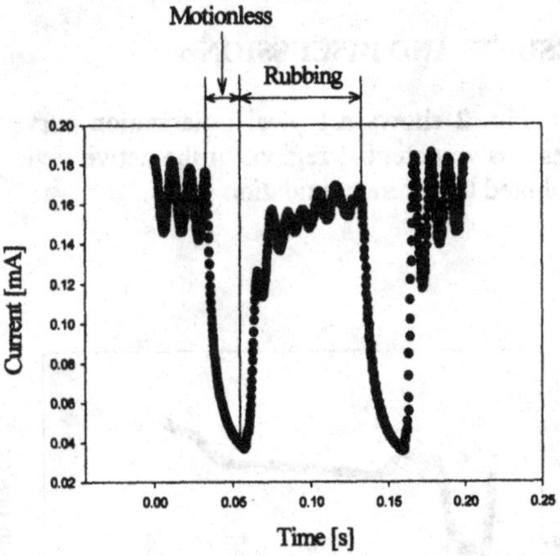

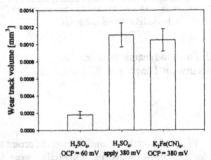

Fig. 4: Current recorded during wear experiments for anodic applied potential (+2V) in 0.5 M H_2SO_4 for a sphere motion cycle of 0.2 s

Fig. 5 shows the volume of the wear track in 0.5 M H_2SO_4 solution versus applied (or free) potential. One can observe from the figure that wear is negligible when a cathodic potential (- 0.9 V versus Ag / AgCl) is applied. At open circuit potential (OCP) some wear is measured and at an applied anodic potential the wear track volume is increasing as applied anodic potential is increased. These results suggest that the formation of oxide is indispensable to the wear process and that the wear rate is related to the oxide formation.

During wear experiments on tungsten in $K_3Fe(CN)_6$ at pH 3, the measured open circuit potential was 380 mV. This particular potential was applied during wear experiments in H_2SO_4 (0.5 M)

Fig. 5: Wear track volume versus applied (or free) potential for experiments conducted in 0.5 M H_2SO_4

solution. Fig. 6 shows the wear track volume for wear experiments in H_2SO_4 (0.5 M) at OCP (60 mV), at 380 mV, and in $K_3Fe(CN)_6$ (0.1 M, pH 4) at OCP (380 mV). One can notice that the wear is identical for wear experiments in $K_3Fe(CN)_6$ at OCP (380 mV) and applying an equivalent potential in H_2SO_4. This result suggests that in this system electrochemical polarization can be used to simulate the oxidizing action of a corrosive environment by imposing a controlled potential.

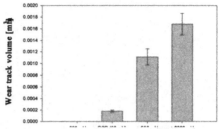

Fig. 6: Wear track volume versus applied (or free) potential for experiments conducted in 0.5 M H_2SO_4 at OCP (60mV), $K_3Fe(CN)_6$ at OCP (380 mV) and H_2SO_4 applying 380 mV

CONCLUSIONS

Potentiodynamic measurements were carried out on tungsten samples, which indicated the formation of a passive film, especially on applying an anodic potential. In addition, tribological experiments carried out under different potentials. The current measurements taken during these experiments, coupled with the wear track depths, confirmed the earlier observation regarding the passivation of the tungsten surface, and also indicated that this passivation is necessary in order for any significant abrasion to take place. These results seem to support the model for tungsten CMP proposed by Kaufman, which involves the continuous formation of an oxide layer and its removal by abrasive particles.

REFERENCES

1. F.B. Kaufman, D.B. Thompson, R.E. Broadie, M.A. Jaso, W.L. Guthrie, D.J. Pearson and M.B. Small, J. Electrochem. Soc. **138**, p. 3460 (1991)

2. E.A. Kneer, C. Raghunath, J.S. Jeon and S. Raghavan, J. Electrochem. Soc. **143**, p. 4095 (1996)

3. V. Brusic, D. Scherber, F. Kaufman, R. Kistler and C.C. Streinz, Abstract 501, *The Electrochemical Society Meeting Abstracts*, V. 96-2, p. 607, San Antonio, TX, Oct. 11-16, 1996

4. E. A. Kneer, C. Raghunath, V. Mathew and S. Raghavan, J. Electrochem. Soc. **144**, p. 3041 (1997)

5. D.J. Stein, D. Hetherington, T. Guilinger, and J.L. Cecchi, J. Electrochem. Soc. **145**, p. 3190 (1998)

EFFECT OF PARTICLE SIZE DURING TUNGSTEN CHEMICAL MECHANICAL POLISHING

MARC BIELMANN, UDAY MAHAJAN and RAJIV K. SINGH*
Department of Materials Science and Engineering and Engineering Research Center for Particle Science and Technology, University of Florida, Gainesville, FL 32611
* rsing@mail.mse.ufl.edu

ABSTRACT

Abrasive particle size plays a critical role in controlling the polishing rate and the surface roughness during chemical mechanical polishing (CMP) of interconnect materials during semiconductor processing. Earlier reports on the effect of particle size on polishing of silica show contradictory conclusions. We have conducted controlled measurements to determine the effect of alumina particle size during polishing of tungsten. Alumina particles of similar phase and shape with size varying from 0.1 μm to 10 μm diameter have been used in these experiments. The polishing experiments showed that the local roughness of the polished tungsten surfaces was insensitive to alumina particle size. The tungsten removal rate was found to increase with decreasing particle size and increased solids loading. These results suggest that the removal rate mechanism is not a scratching type process, but may be related to the contact surface area between particles and polished surface controlling the reaction rate. The concept developed in our work showing that the removal rate is controlled by the contact surface area between particles and polished surface is in agreement with the different explanations for tungsten removal.

INTRODUCTION

Chemical mechanical polishing (CMP) has been widely accepted for oxide and metal layer planarization to eliminate step coverage concerns and improve lithographic resolution. Tungsten (W) CMP is used to remove the excess W deposited by nonselective chemical vapor deposition (CVD) for the formation of vias in IC multilevel interconnects.

Regarding the mechanism of Tungsten CMP, Kaufman et al. [1] proposed a model based on the sequential chemical formation of the passivating layer and its mechanical abrasive removal. Recently Stein et al. [2] reported results based on electrochemical measurements during polishing which suggest that the removal mechanism of W during CMP does not require a blanket passive film or the oxidation of all the removed tungsten, which contradicts the previous model.

According to the empirical Preston equation [3] described as follows

$$RR = K_pPV \qquad [1]$$

the polish rate varies linearly with pressure (P) and linear speed at the wafer-to-pad interface (V). However, the theoretical value of the Preston coefficient, described by Brown et al. [4] as $K_p = 1/2E$ where E is the Young's modulus of the polished surface, does not explain the polish rate variation with other important process variables such as pad properties, slurry chemicals and slurry abrasive.

103

The effect of slurry abrasive size on polish rate is not very clear. Different results have been reported for oxide polishing showing contradicting conclusions. Results obtained by Jairath et al. [5] suggested that the oxide polish rate increases with both abrasive particle size and concentration. However, other reports found that glass polish rate is constant with abrasive size [6], [7] or even decreases with abrasive size [8].

To the best of our knowledge, the study of alumina particle size effect on tungsten polishing rate has not been presented in the open literature. In this paper, we have conducted W CMP experiments using particle sizes ranging from 0.1 to 10 μm to investigate the alumina abrasive size effects on the polish rate. For each size range, the solids loading has also been varied from 2 to 15 weight % to study the dependence on solids loading.

EXPERIMENT

Alumina particles of different size distributions provided by Sumitomo Chemical Inc. were used for the polishing experiments. A Honeywell Microtrac® UPA 150 dynamic light scattering particle size analyzer was used to measure the particle size distribution. Transmission electronic microscopy and scanning electron microscopy were also used to determine the shapes and size of the alumina particles.

All polishing experiments were done on p-type silicon wafers with a titanium nitride (TiN) adhesion layer and a 0.6 μm tungsten (W) film deposited by chemical vapor deposition (CVD). Slurries used for polishing contained from 2 to 15 weight % alumina particles and 0.1 M $K_3Fe(CN)_6$. The pH was adjusted to 4 using nitric acid (HNO_3). A Struers® Rotopol 31 polisher operating at 150 rpm (linear velocity = 275 ft/min) and a pressure of 6.5 psi was used with perforated IC-1000/SUBA IV (Rodel®) stacked polishing pads. During the polishing experiments, the slurry was stirred and circulated with a peristaltic pump with a flow rate of 0.1 l/min. Pad conditioning was performed using grid-abrade diamond pad conditioners (TBW®). Tungsten film thickness was determined by sheet resistance measurements using four-point probe technique. Film thickness was also determined by using SEM. Atomic force microscopy (Digital Instruments Nanoscope III) was used to characterize the surface roughness of samples.

RESULTS AND DISCUSSION

The alumina particles used in these experiments were primarily alpha crystal phase and possessed an approximately spherical morphology as shown in Fig. 1. This figure shows that even though the sizes of the particles vary by nearly two orders of magnitude, the shape is approximately the same. Fig. 2 shows the different particle size distributions that have been used for the polishing experiments. The size distributions match the microscopy data quite well.

Fig. 3 shows the roughness of the polished tungsten surfaces as a function of particle size in the slurry. The measurements were conducted after a removal of approximately 2000 Å of tungsten, using solids loading in the slurry of 5 weight %. The surface roughness was measured from an area of 5 μm x 5 μm using AFM. Fig. 3 shows that the surface finish of tungsten surfaces is independent of the particle size. No scratches have been observed on this tungsten polished surface, except for surfaces polished with 10 – 13 μm particle size where very few shallow (~ 5 nm) scratches have been detected. The fact that the roughness does not increase

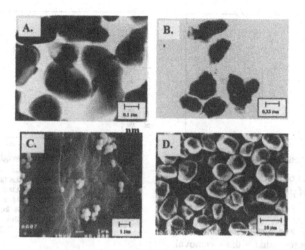

Fig. 1: Micrographs of alumina particles, A. TEM of alumina particle type AKP 30, B. TEM of alumina particle type AKP 15, C. SEM of alumina particle type AA07, D. SEM of alumina particle type AA10.

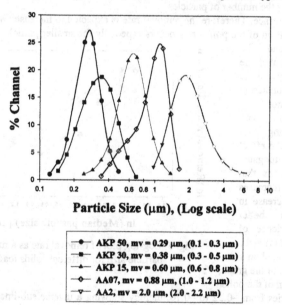

- AKP 50, mv = 0.29 μm, (0.1 - 0.3 μm)
- AKP 30, mv = 0.38 μm, (0.3 - 0.5 μm)
- AKP 15, mv = 0.60 μm, (0.6 - 0.8 μm)
- AA07, mv = 0.88 μm, (1.0 - 1.2 μm)
- AA2, mv = 2.0 μm, (2.0 - 2.2 μm)

Fig. 2: Different Particle Size Distributions measured using Dynamic Light Scattering

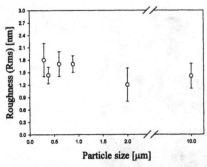

Fig. 3: Roughness of polished samples as a
function of particle size

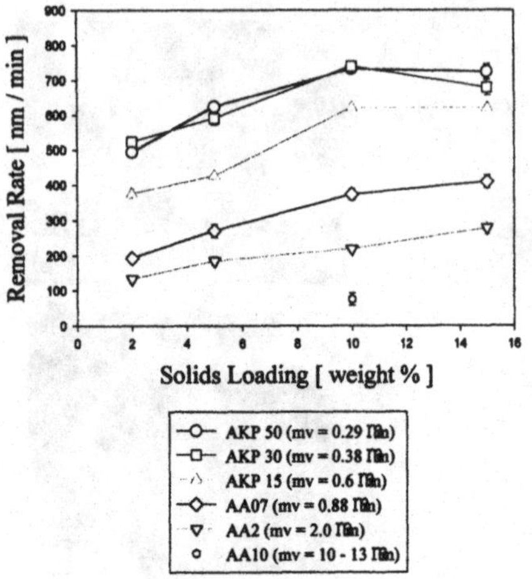

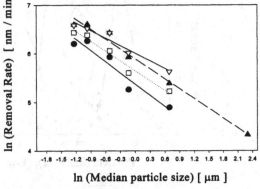

Fig. 4: Tungsten removal rate as a function of
solids loading for different particle sizes

with particle size and that no scratches are observed suggests that the removal mechanism is not a scratching type process. Fig. 4 shows that the tungsten polish rate increases with increased solids loading. For a given particle size, increasing the solids loading will lead to an increase the number of particles interacting with the surface. Therefore the removal rate is expected to increase. We can observe on Fig. 4 that a saturation of the polish rate occurs especially for smaller particle based slurries. This saturation can be explained by the fact that the fill factor is near unity and the number of particles contacting the surface cannot be increased anymore with increasing solids loading.

Fig. 5 shows the effect of particle size on the tungsten polishing rate. It shows that the polishing rate increases substantially with decrease in particle size. To better ascertain the dependence of the polish rate on particle size, the parameters are plotted on a ln-ln scale. The slope of the graph, which is an indication of the polish rate dependence, varies from −0.52 to −0.76, thereby showing a inverse sub-linear dependence of polish rate with particle size.

Fig. 5: Log-log plot of removal rate as a function of
particle size for different solids loadings

This inverse dependence of tungsten removal rate on particle size and the fact that the roughness of the polished surface is independent of particle size indicates that the polishing process is not controlled by an indentation and scratching mechanism, but more by a process limited by the contact area between the abrasives and the polished surface. As the particle size decreases, the surface area per unit mass increases linearly. However with increased surface area, the contact depth will decrease, because of the lower local pressure. This increased contact area is expected to enhance the kinetics of the chemical reaction by assisting in the transport of passivating species to and products from the polished surface and/or removal of the passivating layer from the W surface. A model explaining these effects will be discussed in detail in a further publication. This model is in agreement with the different explanations for tungsten removal mechanisms, such as Kaufman's model [1] or Stein et al.[2].

CONCLUSIONS

The effect of alumina particle size during chemical mechanical polishing of tungsten has been investigated. No significant change in the surface roughness of tungsten was observed for increasing particle size. The tungsten removal rate was found to increase with decreasing particle size and increasing solids loading. Based on these experimental results we have concluded that the removal rate mechanism is not a scratching type process, but that the removal rate is related to the contact surface area between particles and polished surface controlling the reaction rate.

ACKNOWLEDGEMENTS

We would like to acknowledge Dr. Rajan Nagabhusnam from Motorola for providing the samples used for this study. Part of this research is sponsored by the Engineering Research Center (ERC) for Particle Science and Technology at the University of Florida under the National Science Foundation grant No 94-02989.

REFERENCES

1. F. B. Kaufman, D.B. Thompson, R. E. Broadie, M. A. Jaso, W. L. Guthrie, D. J. Pearson, and M. B. Small, J. Electrochem. Soc. **138**, 3460 (1991)

2. D.J. Stein, D. Hetherington, T. Guilinger, and J.L. Cecchi, J. Electrochem. Soc. **145**, 3190 (1998)

3. F. Preston, J. Soc. Glass Technol. **11**, 247 (1927)

4. N.J. Brown, P.C. Baker, and R.T. Maney, Proc. SPIE **306**, 42 (1981)

5. R. Jairath, M. Desai, M. Stell, R. Tolles, and D. Scherber-Brewer, in *Advanced Metallization for Devices and Circuits-Science, Technology and Manufacturability*, edited by S.P Murarka, A. Katz, K.N. Tu and K. Maex (Mater. Res. Soc. Proc. **337**, 121 (1994)

6. L.M. Cook, J. Non-cryst. Solids **120**, 152 (1990)

7. S. Sivaram, M. H.M. Bath, E. Lee, R. Leggett, and R. Tolles, Proc. SRC Topical Research Conference on Chem-Mechanical Polishing for Planarization, SRC, Research Triangle Park, NC (1992), proc. Vol. #P92008

8. T. Izumitani, in *Treatise on Materials Science and Technology*, eds. M. Tomozawa and R. Doremus, Academic Press, New York, (1979), p. 115

9. J. M. Steigerwald, S. P. Murarka and R. J. Gutmann, *Chemical Mechanical Planarization of Microelectronic Materials*, John Wiley & Sons, Inc. (1997)

TUNGSTEN CHEMICAL MECHANICAL POLISHING ENDPOINT DETECTION

Larry Sue, Jörn Lützen, Simon Gonzales
Motorola, Inc. 2200 W. Broadway Rd., Mesa, AZ 85202
Fred Wertsching, Reza Golzarian
Luxtron Corporation, 2775 Northwestern Parkway, Santa Clara, CA 95051

ABSTRACT

Conventional methods of controlling tungsten Chemical Mechanical Polishing (WCMP) processes have included blanket wafer polish rate and platen temperature monitoring. In order to enhance process control of WCMP, a second generation motor-current sense endpoint system was chosen as the evaluation tool. The endpoint detection system (Luxtron 9300) was installed on a WCMP polisher (IPEC 472) at the Motorola MOS6 facility. Endpoint algorithms were evaluated under various polishing conditions during time periods between pad changes. The data collected over 3 months led to a reliable endpoint recipe for both the contact and via WCMP process. The endpoint data compared with the conventional timed recipe, indicated that a time reduction at WCMP could be achieved, with a 10% increase in throughput. In addition, contact and via W-plug endpoint data showed a strong signal versus pattern density for given device types. In summary, the motor current endpoint system tested verified reliable endpoint, process control, and characterization capabilities.

INTRODUCTION

Polish time fluctuations are common occurrences in any Chemical Mechanical Planarization (CMP) process. CMP removal rates are affected by many factors such as pattern density, slurry condition, pad condition and polishing parameters such as down pressures and rotation rates. Historical CMP recipes have been based on time with separate recipes for contact and via layers. The Luxtron Real-Time Controller (LRTC) system provided the capability to use endpoint methodology to manage end-of-process versus fixed time recipes.

The system controller is designed to process and analyze signals that are relevant to the change in the tungsten CMP process and equipment. Implementation of active analog gain, anti-alliance and anti-harmonic filtering in combination with higher than 12 bit analog to digital data acquisition methodology enhanced the signal to noise ratio. Furthermore, using an Automatic Gain Control (AGC) and signal normalization, allowed optimization to yield improved signal differential detection.

Numerous combinations of signal averaging to smooth out the incoming motor current noises were evaluated. Initial endpoint recipe development did not result in good repeatability. The recipe would endpoint on several wafers, with the remainder defaulting past the endpoint set time. During the interpretation of the curve traces it was determined that a portion of the trace behavior was characteristic of the polisher and not the wafer film material.

In principle, the endpoint methodology considers that as a particular layer clears, the material on the next layer exhibits a different coefficient of friction (change in the current due to a difference of frictional loading). However another factor affecting the curve traces was found to be the carrier arm movement motion across the primary platen and the conditioning head (both

109

move back and forth across the radius of the polishing pad). The load that the carrier itself and the conditioning head impose on the primary platen was found to have an effect on the incoming signal, resulting in an oscillation of the curve traces. It was therefore necessary to remove the effects of these oscillations through the application of data averaging. This modification allowed improved synchronization with the periodic behavior of the carrier head and the conditioning transverse motor. Various experiments were carried-out with this modification relative to the detection criteria. Extensive off-line reprocessing of passive data was evaluated in order to optimize the endpoint recipe prior to final testing on production material.

During a three month period the following was evaluated: 1. Repeatability of the endpoint process, 2. Endpoint methodology versus CMP process and 3. Electrical analysis on device material.

EXPERIMENT

The traces below illustrate unconditioned and conditioned data with endpoint. The trace as shown in Figure 1 represents the raw current signal from the motor. By applying gain control and averaging to the raw data, a conditioned trace results as shown in Figure 2. Both single and double peaks were observed on given material. In order to endpoint on the second peak, system parameters were defined to look for endpoint after a designated amount of time had transpired (see Figure 2). Note: Standard timed condition is 3.0 minutes versus 2.0 minutes with endpoint.

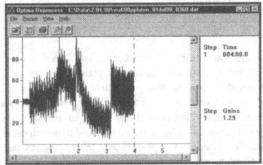

Figure 1 - Raw Motor Current Data - Unconditioned Trace

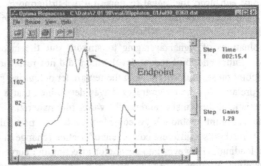

Figure 2 - Conditioned Motor Current Trace with Endpoint

Factors which effect endpoint, include compatibility of endpoint detection criteria with process signature curve and polishing process variations. The process signature curve will changes with variations in the process. Variability exists as a result of film conditions, polishing variation, and differences in via or contact density layouts. The CMP/LRTC recipe was designed to minimize the induced variations in order to provide stable endpoint conditions. As noted, if there is an incompatibility between the endpoint recipe and the process, an endpoint may not occur at all. Process variations include slurry and pad conditions that can have a significant effect on endpoint. These differences can be a result of barrel-to-barrel variation, mixing effects, concentration gradients (change in particle size distribution) and polishing pad surface wear.

DATA COLLECTION AND ANALYSIS

Passive data was collected over a given period and plotted in a Multi-Vari chart (Figure 3) to observe the distributions over time.

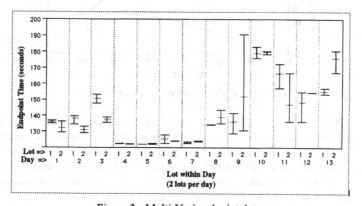

Figure 3 - Multi-Vari endpoint data

The data on lot 2 of day 9 showed a wide process variation (endpoint time increase) from the previous set of data. An analysis of the CMP status log with the endpoint data indicated that slurry was not being pumped to the pads on the tool. The problem was not detected until two hours after it occurred. The root cause was traced to software issues in the slurry distribution system. Note: Real-time endpoint control with set specification limits would have resulted in a warning at the time of excursion. Observation over a three day period examining two lots per day and three wafers from each lot (Figure 4), showed that half of the total variation is explained by lot differences. Wafer-wafer variation within a lot is significantly smaller than between lots.

Figure 5 illustrates endpoint time versus pad wear. In order to eliminate the effect of topological density differences, pad life wear versus endpoint time was examined on two different device types. Note: Typical pad changes occur after a wafer count of 400.

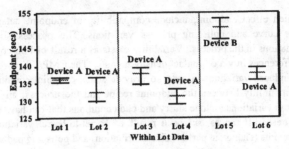

Figure 4 - Lot Variation Analysis

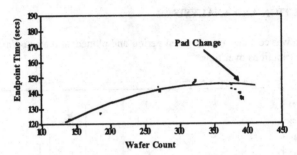

Figure 5 - Wafer Count versus Endpoint Time

No significant decrease in endpoint time after the pad change was observed. Additional analysis would be necessary to determine why the new pad did not return to the lowest endpoint time and if the pad-life could be extended. In production, a guard-band window (increase in endpoint time) could be used to determine pad changes versus wafer count. Endpoint data would facilitate optimizing the trade-off between extended pad life usage (uniformity dependent) and cycle time (polish time increase).

A review of the endpoint times from four production lots prior to slurry barrel change and after is plotted in Figure 6.

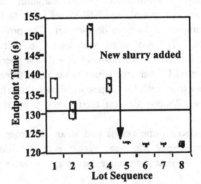

Figure 6 - Slurry Change Versus Endpoint

The chart shows those endpoint times after slurry refill are shorter and more consistent than before the refill. Though the actual failure mechanism was not determined, the barrel-to-barrel slurry variation or mixing effect could result in the different endpoint behavior. Note: A pad change occurred after the slurry change, thus not a contributor to the overall effect.

FULL FLOW ELECTRICAL DATA

Evaluation lots at via and contacts were processed using the WCMP LRTC endpoint system. Both lots were tested under the following conditions: 1. Top of curve endpoint (refer to Figure 2), 2. Top of curve endpoint + 20% overpolish and 3. Termination after current standard recipe (timed).

The data collected at contact (sample data Figure 7 - 8) was compared with similar device types for contact resistance and comb leakage (Note: test lot circled). All the data for each split was combined in order to compare the distribution in reference to other lots. As shown in Figure 7-8, the data splits did not have a large effect on the overall electrical process control results.

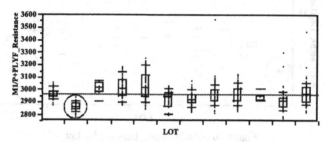

Figure 7 - M1/P+PLYF_Resistance By Lot

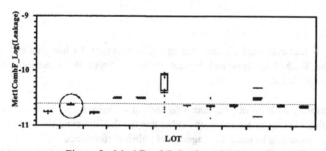

Figure 8 - Met1CombF_Leakage By Lot

The data collected at via (Figure 9-10) were compared with similar device types for via resistance and comb leakage (Note: test lot circled). All the data for each split were combined in order to compare the distribution in reference to other lots. As shown in Figure 9-10, the data splits did not have a large effect on the overall electrical process control results. Thus, the initial test of the endpoint tool at contact and via did not have an impact on the electrical process

113

control test structures. Cross sections of both the via and contact splits using endpoint showed similar oxide recess to standard material (600 to 800 Å).

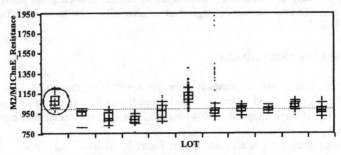

Figure 9 -M2/M1ChnE_Resistance By Lot

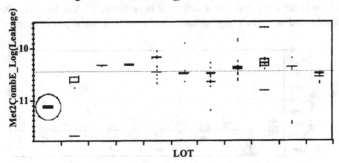

Figure 10 - Met2CombE_Leakage By Lot

SUMMARY

This study that evaluated a motor current endpoint system for tungsten and tested multiple aspects of the WCMP process and system control. From the evaluation performed, the following was concluded:

• Good repeatability with-in-lot (less than 4 seconds of variation)
• Endpoint adjusts the lot-to-lot variations either due the the film conditions or pattern density
• The endpoint traces can be used for diagnosis of polish performance
• Reduced cycle-time with endpoint versus the use of conventional timed recipes (time reduction)

In summary the use of motor-current endpoint versus timed WCMP recipes provides improved process control for production worthy applications.

CHARACTERIZATION OF SLURRY SYSTEM AND SUPPRESSION OF OXIDE EROSION IN ALUMINUM CMP (CHEMICAL-MECHANICAL PLANARIZATION)

Lei Zhong, Jerry Yang, Karey Holland, Joost Grillaert*, Katia Devriend*, Nancy Heylen* and Marc Meuris*

IPEC, 4717 E. Hilton Ave., Phoenix, AZ8503,   lzhong@ipec.com

*IMEC, Kapeldreef 75, B-3001 Leuven, Belgium

ABSTRACT

In this work, we investigated the dependence of the removal rate upon the oxidizer (peroxide) addition into commercially available slurries for a variety of films such as aluminum, titanium, titanium nitride and oxide. We found that the barrier layer materials were extremely sensitive to the peroxide addition while the removal rate varied only slightly for aluminum and oxide. The selectivity to titanium and titanium nitride drops from as high as 1000 to almost close to 1 as the mixture ratio (peroxide : slurry) increases. We proposed that the barrier layer be used to protect the oxide from being over-exposed and suppress the erosion eventually. This can be easily realized by dividing the process into two steps with each step being run at a specific peroxide mixture ratio. The experimental result unambiguously proved, for the first time, the effectiveness of this approach.

INTRODUCTION

Peroxide is suggested by many vendors as the oxidizer for metal polish and recommended as one component of their slurry system. It is still unclear, however, what roles the peroxide plays in the polish process. The original purpose of this work is to evaluate the impact of this variable upon the process performance. The experiment result turns out to be inspiring and suggests a new approach to minimization of the oxide erosion.

EXPERIMENT RESULTS AND DISCUSSION

We chose a commercially available slurry, *SA* as an example and investigated the dependence of removal rate upon the peroxide addition. The result is shown in Fig.1. All the data points linked with lines (solid or dotted) have been collected on the identical sheet wafers polished with the same recipe, except the slurry composition. The non-uniformity is below 10% in most cases except for the barrier layer films polished without peroxide added, which indicates that the misrepresentation of removal rate which is the average of 49-point polar map, is negligible. As may be seen, the aluminum removal rate is a very weak function of the mixture ratio. This is a strong evidence that the chemicals in the slurry are capable of oxidizing aluminum so that peroxide is not indispensable for aluminum polishing at all. A closer examination shows that the removal rate drops slightly but linearly with the peroxide addition. A plausible explanation for this phenomenon is the dilution of abrasive particles. If the chemical reaction (oxidation) is fast enough, the removal rate must be mechanically dominated and roughly proportional to the availability of abrasive particles in contact with the wafer surface[1]. This is reasonably in agreement with our result shown in Fig.1. Also independent of the peroxide addition is the oxide removal rate which is very low, almost two orders of magnitude lower than that of aluminum. It is worthy noting that the selectivity to the oxide (1:100) is generally considered acceptable and there is practically little room for further improvement.

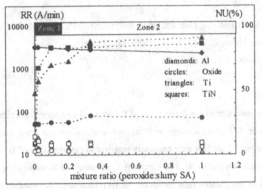

Fig.1. peroxide-slurry *SA* system. The peroxide is 31% solution. The lines are drawn to guide the eyes only. See the context for details.

What makes Fig.1 interesting is the removal rate of barrier layer films, which exhibit a strong dependence upon the peroxide addition. As might be seen, titanium polishes at least one order of magnitude slower than aluminum if no peroxide is added into the slurry. The removal rate increases exponentially with the peroxide addition and does not level off at a value slightly above that of aluminum until the mixture ratio reaches at 0.3 (1:3). Titanium nitride seems even more sensitive to the presence of peroxide. The removal rate is extremely low (less than one thousandth of aluminum's) in the as-received slurry, which partly explains the abnormally high non-uniformity we noted earlier because the relative fluctuation could be extremely large even though the absolute deviation is very small. Notice that titanium nitride is even harder to polish than the oxide. With addition of as little as 0.1 (1:10) peroxide, however, the removal rate of titanium nitride soars up to the level of Al and becomes rather stable afterwards.

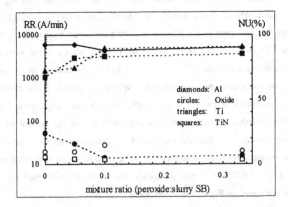

Fig.2. peroxide-slurry *SB* system. The peroxide is 31% solution. The lines are drawn to guide the eyes only. See the context for details.

116

Although peroxide is employed in many commercially available slurries as the oxidizer, the response of the removal rate of the barrier layer materials to the peroxide addition may vary dramatically from one product to another. Shown in Fig.2 is the results collected for another slurry *SY*. As one may see, the dependence of the barrier layer removal rate upon the peroxide addition exhibits a similar trend as demonstrated by slurry *SA* in Fig.1. In spite of this, however, slurry *SB* gives rise to an appreciable removal rate for the barrier layer even without any peroxide addition, which distinguished itself from slurry *SA*.

In terms of the selectivity to the barrier layers, the peroxide-slurry *SA* system can be characterized by its mixture ratio into two zones, as shown in Fig.1 which bears more or less resemblance to the phase diagram widely used in thermodynamics. Zone 1 refers to a system with the mixture ratio below 0.3. The selectivity is greatly larger than 1 and is mixture ratio dependent. While in Zone 2, the selectivity is independent of the mixture ratio and is almost equal to 1. Traditionally, the system is recommended to be used in Zone 2 with a fixed mixture ratio of 0.3 (1:3) in order to make the removal rate and consequently the process as stable as possible.

We propose that the system characters be utilized to tailor and control the process. Our practice is to divide the process into two steps, called polishing and clearing, respectively. Each one of the steps is designed to perform a specific function and is carried out in a designated zone. The polishing step is to remove the aluminum layer with the slurry mixed in Zone 1, followed by the clearing step which is run in Zone 2 to get rid of all the metal residuals on the field.

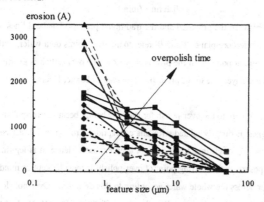

Fig.3. oxide erosion versus feature size obtained from a preliminary experiment. The data collected on the same wafer but different dies is represented by the same symbols linked with a line. The erosion obviously increases with the over-polish time despite of the data scatter.

This approach offers unprecedented advantage in suppressing the oxide erosion. Since the erosion is generally believed to be critical to the yield management its importance to the success of CMP application could never be over-emphasized. Oxide erosion occurs because of the limited selectivity to the field oxide (chemical effect) and the metal recess (dishing) in the trench (geometrical effect). Even with an acceptable selectivity of 100:1, the erosion can be amazingly large depending on the feature size and the over-polish time. This is illustrated by Fig.3 which was obtained from our previous study. This figure can be explained qualitatively by the model proposed by Rentl, who discussed the step height reduction (SHR) in terms of the elasticity theory[2]. Other workers did similar theoretical treatment.[3,4] Apparently, any erosion model should be claimed successful only if it can be reduced to the SHR model when the selectivity is one. Following modifications should be considered in order to explain the erosion. First, the chemical effect is expectedly playing a critical role in erosion since it occurs due to the presence of two phases. Second, the Poisson's ratio should not be ignored because the lateral deformation could be quite significant sometimes. A solution to the differential equations with non-zero Poisson's ratio is thus indispensable. Third, the SHR model has to be extended to include the dynamic issue (over-polish).

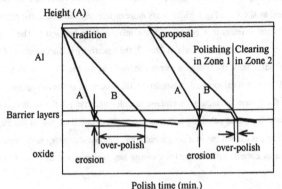

Fig.4. a schematic comparison between the proposed and the traditional process. The bold lines represent the film height measured from any given reference plane. A and B refer to any two points on a wafer, with the assumption that removal rate sees its minimum at point B. In the proposed process, the over-polish is greatly reduced because the onset of the oxide exposure is delayed due to the protection provided by barrier layers.

We restrict our discussion here to the over-polish for the time being, because over-polish does not exist in SHR and the approach we suggest is directly related with this conception. Erosion is a direct consequence of the over-polish because the oxide would be left intact if it had not been exposed. Hence, it is legitimate to define the very beginning of the oxide exposure as the starting point of over-polish. It must be kept in mind that over-polish varies from one point to another across the whole wafer according to this definition. Over-polish and under-polish may coexist depending on the feature dimension as well as the non-uniformity. In order to clear the metals on the wafer scale (just-polish) a minimum over-polish is required for the rest of the wafer area as compared with the point which is exposed last. It is the minimum required over-polish that determines the oxide erosion. The method we proposed is basically to utilize the barrier layer to keep the oxide from being over-exposed. This is schematically illustrated in Fig.4, where A and B represent any two points on a wafer. Assume that for any reason the removal rate is higher at point A than at point B, so that when the oxide is exposed at point A aluminum is still left on point B. In the traditional process, By the time the metals are cleared at point B, the wafer has experienced a severe over-polish and would see a significant oxide erosion at point A. This occurs because the removal rate of the barrier layers is comparable to that of aluminum. In the process proposed, the removal rate is greatly retarded at point A when the barrier layer is encountered so that the oxide exposure is substantially delayed.

118

In other words, the oxide is protected by the barrier layer until aluminum is completely removed at point B. The barrier layer is then cleared by switching the slurry from Zone 1 to Zone 2. By this way, the over-polish at point A is reduced and the erosion is suppressed.

The implementation of the process proposed demands the ability to change the slurry composition swiftly. This can be easily realized on Avantgaard-676 polisher because the polish platen is so compact that the time needed to prime the pad is relatively short. The procedure we used in this work was to quench the process when the barrier layers were completely exposed, switch the slurry from Zone 1 to Zone 2 and then clear the barrier layers. The process was completed on one platen with the wafer being left away from the pad in between the two steps. The result of a comparison experiment is shown in Fig.5, where two wafers were polished with the traditional process and the proposed process, respectively. The polishing time was the same in both cases. The over-polishing time, however, is much longer in the later case since the removal rate of aluminum is much faster when a slurry in zone 1 is used, according to Fig.1. This is evidenced by the deeper metal dishing in Fig.5 (b). Nevertheless, the oxide erosion is less than half of that obtained with traditional process as shown in Fig.5.(a).

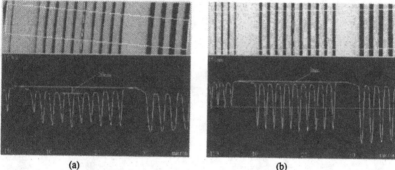

(a)   (b)

Fig.5. 40x10 μm AFM images of a trench array (S/L=1/0.7) after polished with the traditional process (a) and the process proposed (b), respectively. Although the total polish time is the same in both cases, the metal dishing is almost doubled in (b) since the removal rate is much higher due to the use of the slurry in Zone 1. Nevertheless, the oxide erosion is much smaller.

It should be pointed out that the effectiveness of this approach is heavily dependent upon the characters of the slurry being used. If the removal rate of the barrier is not sensitive to the peroxide addition, like slurry SB for example, the barrier layer cannot provide enough resistance to function as a protection (stop) layer for the underlying oxide even though one might define two different zones for the slurry system. On the other hand, if the slurry exhibits the characters desired, we believe that the same method is applicable to the other metals as well, such as tungsten and copper as long as the barrier layer can be utilized to protect the oxide.

CONCLUSIONS

We examined the impact of the peroxide addition upon the polishing performance. In contrast with aluminum, the barrier layer materials titanium and titanium nitride are extremely sensitive to the peroxide. It is suggested that the barrier layer be used as a stop layer to protect the underlying oxide from being over-exposed so that the oxide erosion can be suppressed. This proposal is bolstered by the experiment.

ACKNOWLEDGEMENT

This work is part of the joint development project between IPEC and IMEC. L.Z. wishes to thank all his colleagues in IMEC for their assistance and hospitality during his stay in Belgium.

REFERENCES

1. L.M.Cook, J.Non-Crystalline Solids, **120**, p.152 (1990).

2. P.Renteln, ULSI-X, (Mater. Res. Soc. Proc. 359, Pittsburgh, PA1995), p.153.

3. N.Elbel, B.Neureither, J.Muller, and B.Ebersberger, CMP-MIC, (1997), p75.

4. N.Elbel, B.Neureither, B.Ebersberger, and P.Lahnor, J. Electrochem. Soc. **145**, p.1659 (1998).

Part IV

Copper Polishing and Related Issues

ROLE OF FILM HARDNESS ON THE POLISH RATES OF METAL THIN FILMS

S. Ramarajan [a], Y. Li [b], M. Hariharaputhiran [a], Y.S. Her [c], and S.V. Babu [a]
Department of Chemical [a] and Mechanical [b] Engineering, Clarkson University, Potsdam, NY.
[c] Ferro Corporation, Penn Yan, NY.

ABSTRACT

Nanoindentation techniques were used to determine the hardness of Cu, Ta & W metal discs and thin films on silicon substrates as a function of load or indentation depth. Cu films exposed to oxidizing solutions containing H_2O_2 exhibited a higher hardness at the surface while no such change was observed for W exposed to ferric nitrate. The implication of these measurements and their relationship to chemical-mechanical polishing rates are discussed.

INTRODUCTION

Planarization of metal films like copper, tantalum and tungsten appears to be dominated by the abrasion of the chemically modified surface layer of these films during chemical-mechanical polishing (CMP). The surface modification can be of different forms, depending on the film and the chemicals present in the polishing slurry. Thus in case of copper, a H_2O_2/glycine slurry oxidizes[1] the film while ferric nitrate simply dissolves it[2]. Addition of benzotriazole (BTA) to the slurry will lead to the formation of a passivating polymeric Cu-BTA film[3]. Hence, the mechanical and chemical characteristics required of the abrasive particles to obtain an optimal polish process in each of these slurries can be very different. These requirements are likely to be determined by the mechanical properties like hardness, elastic modulus etc., of the surface modified layers of the films as well as the effect of the slurry chemicals, if any, on the abrasive particles themselves. Several theoretical investigations[4,5] of the CMP process, based on Hertzian indentation model, also confirm the important role of these parameters. Indeed, Tseng et al.[6], reported an inverse linear relationship between the surface hardness and the removal rate of oxide film. It is certain that the mechanical properties of these modified surface layers will be very different from those of the same materials in bulk form. Since the surface modified layer is likely to be only a few nanometers thick, nanoindentation techniques are ideally suited for measuring the relevant mechanical properties of these surface layers.

In our previous work[2], we investigated the effect of alumina abrasive particulate properties, such as porosity and crystalline morphology, on the metal polish rates. We reported an increase in the polish rate with decrease in the particle porosity and an increase in the α-content of the alumina particles, both of which result in an increase in particle hardness. Hence, it was inferred that an increase in polish rate was caused by the increase in particle hardness. This suggestion was also supported in general terms by the measured polish rates of W, Ta, and SiO_2 in DI water, even though only Ta exhibited a similar threshold behavior. However, since the hardness of the particles or of the films was not available, a relationship between the hardness and polish rate could not be determined. In this paper we take the first steps to establish such a relationship. The main goal of this work is to measure the hardness of metallic (Cu, Ta and W) thin films deposited on silicon substrates and metallic discs using nanoindentaion techniques[7]. These measurements were made in several different chemical environments of relevance to CMP and it was found that the hardness values varied as a function of depth from the surface.

Mat. Res. Soc. Symp. Proc. Vol. 566 © 2000 Materials Research Society

EXPERIMENTS

3 μm thick copper films and 1 μm thick tantalum films were sputter-deposited on 125 mm diameter single crystal Si wafers using a TORR International CRC-150 sputtering system. A 50 nm Ta adhesion/barrier layer was deposited prior to copper deposition. 1 μm thick tungsten coated silicon wafers were purchased from Silica source technology.

The hardness of films and metal discs were measured using Nanomechanical test instrument from Hysitron Inc. A Berkovich diamond tip was used in all the measurements. The measurements were made at 10 different peak loads of 300 μN and 600 μN and at intervals of 1000 μN between 1000 and 8000μN. The loading and unloading were done at a constant rate of 250 μN/s, except at the two lower loads where a rate of 100 μN/s was used. After reaching the peak load, the load was held constant at the peak value for 5 seconds to minimize time dependent plasticity. Figure 1 shows a typical load time sequence. The nanomechanical system provides a continuous measurement of load and penetration depth. Figure 2 shows the load versus displacement curves for the copper- and tungsten-coated wafers. The hardness ($H = P_{max}/A_c$)

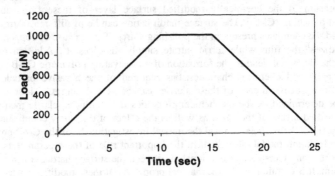

Figure 1. Load-Time sequence - peak load 1000 μN

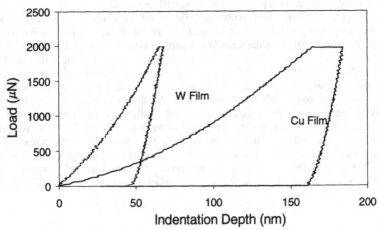

Figure 2. Load vs Displacement curves

124

is computed from the peak load (P_{max}) and contact area A_c. The contact area is obtained from the indenter tip shape function, which relates contact area to the contact depth. Contact depth is the depth up to which the indenter is in contact with the sample at peak load. Hardness measurements were carried out at five different locations on the wafer at the same load. The reported values are obtained by averaging the five measurements.

The effect of chemicals used during the CMP process on the mechanical properties of copper and W thin films was investigated by dipping the wafers coated with these films in 5 wt % H_2O_2 and 0.1 M ferric nitrate solutions, respectively. The wafers were dipped for five minutes, washed in de-ionized water and dried. Hardness of these exposed films was measured as above.

Polishing experiments were carried out on a bench top Struer's polisher with 3 mm thick metal disks (99.999% pure) of cross sectional area 7.5 cm^2. The table speed was 90 rpm and the disk holder was stationary. The applied downward pressure was about 4.3 x 10^4 N m^{-2}. The solids concentration in the slurries was maintained at 3 % by weight and the slurry feed rate was 1 ml s^{-1} for all the experiments. The slurry in the supply tank was stirred continuously with a magnetic stirrer to avoid settling of agglomerated particles. A Suba 500 polish pad was used and was hand conditioned prior to each experiment using a 220 grit sandpaper and a nylon brush.

The polish rate was determined from the difference in the weight of the disk before and after polishing for 3 minutes and the reported values were obtained by averaging over four experiments.

Polishing experiments were also carried out on a Strasbaugh 6CA polisher. The polisher parameters were set at 40 rpm rotational speed of table and quill, 2.76 x 10^4 Nm^{-2} downward pressure and 4 ml s^{-1} slurry flow rate for all the polishing experiments. The slurry in the supply tank was continuously stirred using a magnetic stirrer. The alumina particle loading in the slurry was fixed at 3% by weight.

RESULTS AND DISCUSSION

Figure 3 shows the dependence of hardness on load for as-deposited copper, tantalum and tungsten thin films. An increase in the hardness of copper and tungsten films was observed at lower loads, i.e., for lower penetration depths. This is probably due to the presence of an oxide on the film surface and/or surface defects.

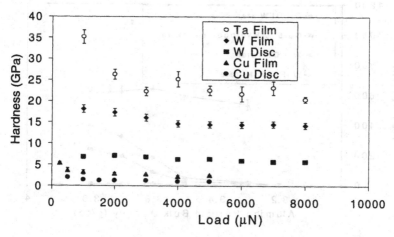

Figure 3. Hardness dependence on load

Earlier we reported a jump in the copper and tantalum polish rates while polishing with alumina abrasives when the abrasive particle density exceeded a threshold value (3.5 g/cc in the case of Cu and 3.6 g/cc in the case of Ta), while no such behavior was observed in the case of W. This sudden increase in the polish rates above a certain threshold particle density was interpreted to be due to an increase in the hardness/elastic modulus of the particles above the hardness/elastic modulus of the materials being polished. This threshold behavior in the polish rates suggests that the hardness values should be highest for W followed by Ta and Cu. This is confirmed only partially since, while both W and Ta films were found to be harder than Cu, a higher hardness was obtained for tantalum films compared to tungsten films. The bulk material hardness values reported in the literature show a lower hardness for tantalum compared to tungsten. The reason for the reverse trend observed for thin film hardness is not clear. One possible reason might be the existence of differences in the hardness values of discs and thin films. The polishing experiments were carried out with metal discs. Hence, hardness values of Cu and W discs were measured and are also shown in figure 3. Interestingly, the hardness of these discs was significantly lower than that of the corresponding thin films. Unfortunately, Ta disc hardness could not be determined, as the surface roughness was high. Thus the above conflict remains unresolved.

We also reported earlier a decrease in the copper (disc and thin films) polish rate in the presence of 5 wt % H_2O_2 compared to the polish rate in de-ionized water, for all alumina particle densities. A set of these results is shown in Figure 4. Electrochemical measurements indicate the formation of a passivating film on the copper surface when exposed to H_2O_2 solutions. Also, Hirabayashi et al.[8] reported a 30nm thick copper (II) oxide film on a copper sample dipped for 10 minutes in a 5wt % hydrogen peroxide solution. These results, combined with the observed decrease in the polish rate, suggest that the oxide film on the copper surface is harder than the copper film itself. This is confirmed by the measured hardness of copper films dipped in 5 wt % H_2O_2. Figure 5 shows the hardness values of the exposed Cu films obtained at various loads along with those of as-deposited Cu films. Higher hardness, at low loads, obtained for the sample dipped in H_2O_2 confirms the hardening of the copper surface, due to the formation of an oxide film. The nearly same hardness values obtained at higher loads, i.e., higher indentation depths, for both is an indication that the oxide film is confined to the film surface, as may be expected.

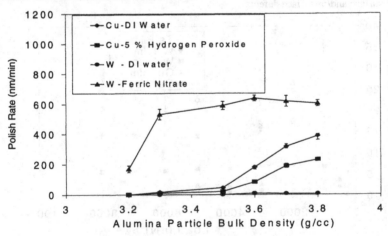

Figure 4. Cu and W polish rates as a function of alumina particles bulk density

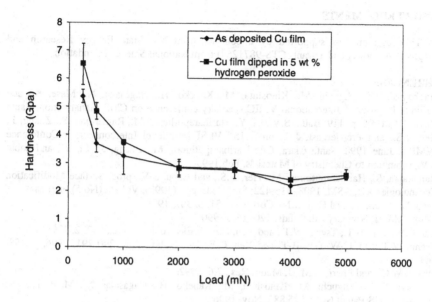

Figure 5.Effect of Hydrogen Peroxide on the Cu film hardness

Investigation of the effect of particulate properties during CMP of W showed a significant increase in the polish rate in the presence of ferric nitrate compared to the polish rate in de-ionized water, at all alumina bulk density values (shown in figure 4). Kaufmann et al.[9], attributed the increase in the polish rate in the presence of ferric nitrate to the "softness" of the passivating oxide film compared to W. Potentiodynamic experiments and open circuit potential measured as a function of time indicate passivation of W surface. However, the hardness values of tungsten films exposed for 5 min to 0.1 M ferric nitrate, even at the lowest load (300μN), were the same as those of as-deposited W films within experimental error. Since a 10 nm indentation depth was observed at the lowest load, it is possible that the thickness of the oxide film is smaller and its effect does not manifest itself on the hardness measurement.

CONCLUSIONS

Hardness measurements of as-deposited Cu film and those exposed to H_2O_2 containing solution, using nanoindentation techniques reveal a harder surface film in the treated copper films. W films were harder but their hardness did not change after exposure to ferric nitrate. Ta films had the highest hardness. Thin films of W &Ta were found to be harder than their corresponding metallic discs. However, the relation between the relative polish rates of these materials and their hardness values remains unresolved.

ACKNOWLEDGEMENTS

This research was supported by a contract from NY State Energy Research and Development Authority and a grant (CTS-9871264) from National Science Foundation.

REFERENCES:

1. Hirabayashi, H., Higuchi, M., Kinoshita, M., Kaneko, H., Hagasaka, N., Mase, K., and Oshima, J., proc. 1st International VMIC specialty conference on CMP planarization, Santa Clara, CA, 1996, p. 119; Babu, S.V., Li, Y., Hariharaputhiran, M., Ramarajan, S., Zhang, J., Her, Y.S., and Prendergast, J.E., proc. 15th VLSI Multilevel Interconnection Conference (VMIC), June 1998, Santa Clara, CA; Hariharaputhiran, M., Zhang, J., Li, Y., and babu, S.V., submitted to Chemistry of Materials, Feb,1999.

2. Ramarajan, S., Hariharaputhiran, M., Her, Y.S., and Babu, S.V., proc. Surface Modification Technologies XII, ASM, 1998, 415-422; Surface Engg., (1999), Vol 15, (No 5), in press.

3. Ling, Y., Guan, Y., and Han, K.N., Corrosion, 51, p. 367, 1995

4. Cook, L.M., J. Non-Crystal. Solids, 120, 152, 1990

5. Liu, C.W., Dai, B.T., Tseng, W.T., and Yeh, C.F., J. Electrochem. Soc., 143, 2, 1996

6. Tseng, W-T, Liu, C-W, Dai, B-T, and Yeh, C-F, Thin Solid Films, 290-291, 1996, p. 458-463

7. Oliver, W.C., and Pharr, G.M., J. Mater. Res., 7,6, 1992.

8. Hirabayashi, H., Higuchi, M., Kinoshita, M., Kaneko, H., Hagasaka, N., Mase, K., and Oshima, J., US Patent No. 5,575,885, Nov. 1996.

9. Kaufman, F.B., Thompson, D.B., Broadie, R.E., Jaso, M.A., Guthrie, W.L., Pearson, D.J., and Small, M.B., J. Electrochem. Soc., 143, 12, 1996.

Mechanism of Cu removal during CMP in H_2O_2-glycine based slurries

M. Hariharaputhiran[a], S. Ramarajan[a], Y.Li[b], S.V. Babu[ac]

Departments of Chemical Engineering[a] and Chemistry[b], Clarkson University
[c]Center for Advanced Materials Processing, Clarkson University, Potsdam, NY

Abstract

Hydroxyl radical generation has been observed during Cu CMP using hydrogen peroxide-glycine based slurries. While the Cu dissolution/polish rates increased with increasing glycine concentration, the copper dissolution rate decreased with increasing peroxide concentration indicating the occurrence of both dissolution and passive film formation during CMP. This is further confirmed by both *in situ* and *ex situ* electrochemical experiments.

Introduction

Copper CMP under highly acidic conditions leads to severe corrosion problems while under alkaline conditions Cu polish rate selectivity with respect to SiO_2 is unfavorable leading to ILD erosion. Thus an intermediate pH condition (3-7) is better for Cu CMP[1]. H_2O_2-glycine (an amino acid) based slurries containing either silica or alumina abrasive particles are one of the more attractive slurries in this pH regime for Cu CMP[2,3,4,5]. We recently reported[3] that the Cu^{2+}-(glycine)$_2$ complex (chelate) catalyzes the decomposition of hydrogen peroxide to yield hydroxyl radicals (*OH) which play a major role in Cu removal when polished using these slurries. This paper will report on the mechanism of Cu removal during CMP in H_2O_2-glycine based slurries. Results from both *in situ* (during polishing) and *ex situ* (rotating disc electrode) electrochemical experiments and polishing experiments are reported. The effect of benzotriazole as a dissolution inhibitor during CMP in these slurries has also been studied.

2. Experiment

2.1 Chemical-Mechanical Polishing

Polishing experiments were performed using a bench-top Struers DAP-V polisher and 3 mm thick copper disks (99.99% pure, Aldrich) with a cross sectional area of 7.5 cm^2. The table speed was set at 90 rpm and the disk holder was held stationary. The applied downward pressure was about 6.3 psi (41.4 KN/m^2). Alumina particles with a mean aggregate size of around 350 nm, prepared and supplied by Ferro Corporation, were used as the abrasives[6]. The solids concentration in the slurries was maintained at 3 % by weight and the slurry feed rate was 1 ml/s for all the experiments. The slurry in the supply tank was stirred continuously with a magnetic stirrer to avoid settling of aggregated particles. Suba 500 was used as the polish pad. The pad was hand conditioned prior to each experiment using a 220 grit sand paper and a nylon brush. The polish rate was determined from the difference in the weights of the disk before and after polishing for at least three minutes and the reported values were obtained by averaging over four experiments

2.2 Copper Dissolution

Dissolution experiments were carried out in a 500 ml glass beaker containing 400 ml of the etchant solution. A rectangular copper coupon (2.3 cm x 2.3 cm x 0.2 cm, 99.99 % pure) was used as the sample. The copper coupon was first washed with dilute HCl to remove any native oxide from the surface, dried in an air stream and weighed. It was then immersed in the solution for a predetermined time interval. The solution was stirred using a mechanical stirrer at 1000 rpm, unless specified otherwise, to minimize mass transfer effects. The coupon was removed, washed repeatedly with DI water, dried in an air stream and reweighed. The weight loss was

129

used to calculate the dissolution rate and the reported rates were obtained by averaging over four experiments, each spanning at least 2 minutes.

For cases where the Cu dissolution is very small (as in the presence of an inhibitor such as BTA), the weight loss method is not sufficiently sensitive for short immersion times. In those cases, electrochemical linear polarization experiments were performed with an EG&G Princeton Applied Research Model 273 potentiostat/galvanostat to obtain the corrosion current I_{corr} which was converted into a dissolution rate using Faraday's law. An EG&G Princeton Applied Research Model 352 SoftCorr TMII corrosion software was used to control the potentiostat/galvanostat. A standard corrosion cell consisting of three electrodes, namely, a counter electrode (platinum), a reference electrode (saturated calomel electrode (SCE)), and a working electrode was used. The details of the corrosion cell and the experimental procedure have been discussed in detail elsewhere[7,8]. A rotating copper disk electrode (3.1 mm in diameter) was used as the working electrode. The reference electrode was inserted into the corrosion cell through a luggin bridge whose tip was 1-2 mm from the working electrode. Saturated K_2SO_4 was used as the luggin solution. The voltage scan rate and range were fixed at 0.1 mV/s and ± 20 mV vs. open circuit potential, respectively. The corrosion current and corrosion potential (E_{corr}) were calculated using Stern-Geary[9] equation.

2.3 *In situ* Electrochemical Measurements

In situ electrochemical measurements (during polishing) were performed using the Struers DAP-V polisher and 3 mm thick copper disks with a cross sectional area of 6.16 cm^2. The polisher set up and the polishing conditions were maintained the same as mentioned earlier in section 2.1. A three electrode set up namely, a working Cu electrode (polishing substrate), a platinum counter electrode and a SCE reference electrode was used. The reference electrode was placed close to the polishing substrate (within 1 inch) and a 0.1 M NaClO$_4$ was used as the supporting electrolyte in all the cases. In addition a slurry build up of at least 2 cm was maintained above the pad in order to have a proper electrical contact between the working electrode and the reference electrode. The same potentiostat/galvanostat as used in the *ex situ* studies (section 2.2) was used.

3. Results and Discussion
3.1 Copper Dissolution and Polishing

Figure 1 shows the copper dissolution rate as a function of rotational speed in the presence of 5 wt % H_2O_2 and 1 wt % glycine. The copper dissolution rate was 79 nm/min under static conditions (0 rpm rotational speed) and increased with increasing rotational speeds attaining a saturation value of around 160 nm/min beyond 800 rpm. Apparently, mass transfer effects are not rate-limiting beyond 800 rpm. Hydroxyl radicals are very active species with a life time in the order of microseconds[10] and hence the diffusion of the *OH from the bulk of the solution to the substrate surface is unlikely. The Cu dissolution rate is thus determined by the *OH concentration at the substrate/solution interface. The *OH generation is catalyzed by Cu^{2+}-(glycine)$_2$ chelate, as had been established elsewhere[3], and the concentration of Cu^{2+}-(glycine)$_2$ chelate at the interface determines the copper dissolution rate. It should be noted, however, that the source of Cu^{2+} ions is the substrate itself, i.e., no external copper salt was added to the solution. The dissolved copper readily forms a chelate with glycine[11] which in turn catalyzes the decomposition of H_2O_2 to yield *OH. This raises the possibility of an increase in the Cu dissolution rate with increasing time, as there could be an accumulation of Cu^{2+} ions. However, the Cu dissolution rate did not vary beyond the experimental error over a one to ten minute range. This indicates that the concentration of Cu^{2+}-(glycine)$_2$ (and hence the concentration of

*OH) at the interface attains a steady state value in less than a minute and remains unaltered by the continued dissolution (at least up to 10 minutes).

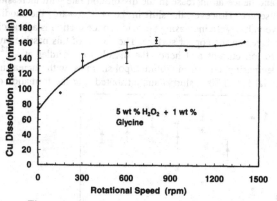

Figure 1: Cu dissolution rate as a function of rotational speed

Figure 2 shows the copper polish rate as a function of glycine concentration in 5 wt % H_2O_2, in the presence and absence of abrasive particles. In both cases, Cu polish rate increases with increasing glycine concentration. Here again, though no copper salt was added to the slurry externally, the copper dissolved/abraded from the sample surface during polishing generates a sufficiently high concentration of Cu^{2+} ions in the vicinity of the copper surface, which complexes with glycine to cataylze *OH generation. Figure 3 shows the copper dissolution/polish rates with added $Cu(NO_3)_2$ in the solution/slurry. As may be expected, the copper dissolution/polish rates increased with increasing $Cu(NO_3)_2$ concentration and leveled off beyond a concentration that, of course, depends on the glycine concentration. Higher the glycine concentration, higher was the Cu^{2+} ion concentration at which the polish rate started leveling off.

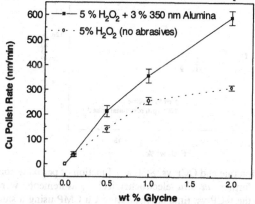

Figure 2: Cu Polish rate as a function of glycine concentration

Figure 4 shows the copper dissolution rate as a function of H_2O_2 concentration in the presence of 1 wt % glycine. The variation in the open-circuit potential (OCP) measured with respect to SCE at 1000 rpm rotational speed, as measured in an *ex situ* electrochemical corrosion cell, is also plotted as a function of H_2O_2 concentration. The copper dissolution rate decreases with increasing peroxide concentration, which is contrary to the expectation. If copper dissolution is controlled by the *OH concentration at the substrate/solution interface, as explained earlier, an increase in the hydrogen peroxide concentration should result in an increase in the *OH concentration and hence an increase in the dissolution rate with increasing peroxide concentration should be observed. However, the shift in the OCP towards the anodic direction, i.e., towards a more positive value, with increasing peroxide concentration is an indication of the formation of a passivating layer on the copper surface. The nature of this passive layer controls the magnitude of the corrosion current and hence the dissolution. Zeidler et al.[12] also made similar observations of decreasing copper dissolution/polish rates with increasing peroxide concentrations (in commercial Rodel 8099 slurry) and attributed it to the formation of a passive film.

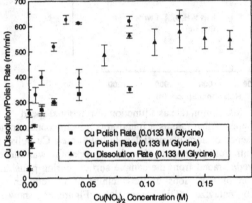

Figure 3: Cu dissolution/polish rates as a function of added copper nitrate concentration

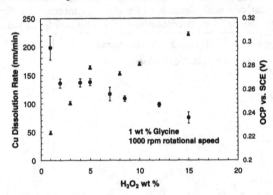

Figure 4: Cu dissolution rate and OCP vs. SCE as a function of peroxide concentration

To investigate it further, *in situ* electrochemical measurements were made during polishing. Figure 5 shows the OCP vs. time curves during Cu CMP using a slurry containing 1 wt % glycine, 5 wt % H_2O_2 and 3 wt % alumina abrasives (3.7 g/cm^3 bulk density), both in the presence and absence of 1 wt % $Cu(NO_3)_2$. The experiments were performed for a period of 10 minutes with polishing stopped after 5 minutes. The OCP remains constant during polishing and once the polishing was stopped, shoots up and stabilizes at a value higher than the OCP obtained during polishing. The increase in the OCP once the polishing was stopped is an indication of repassivation of the copper surface. Also the steep rise in the OCP, once the polishing was stopped, indicates that repassivation occurs rather rapidly within few seconds. Furthermore, with the addition of 1 wt % $Cu(NO_3)_2$ to the slurry, the OCP stabilized at a higher value, when the polishing was stopped, indicating a further increase in the tendency to passivate with the addition of $Cu(NO_3)_2$. Thus copper dissolution and mechanical removal of the passive film are involved during Cu CMP using H_2O_2-glycine based slurries, in addition to the mechanical removal of the copper itself by the abrasives. Further, the occurrence of both dissolution and passivation indicates that either the passive film is porous whose properties determine the dissolution rate or the passivation itself is an intermediate step in the dissolution process as in the case of Cu dissolution in ammonium hydroxide solution[13]. However, such a hypothesis needs more experimental evidence and will be pursued further.

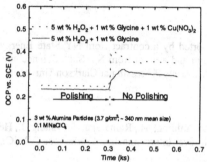

Figure 5: In situ OCP vs. SCE measured as a function of time

3.2 Effect of BTA (inhibitor)

For planarization of uneven surface topography to occur, isotropic etching should be avoided. Higher dissolution rates remove copper from both the protruding as well as the recessed regions, resulting in poor planarization. For a good CMP process, the copper in the recessed regions should be protected from direct dissolution especially if the purely chemical dissolution rate is high as is the case with the H_2O_2-glycine based slurries. In order to prevent the direct etching of copper in the recessed regions, a corrosion inhibitor could be added that forms a non-porous passive film and hence minimizes direct dissolution. An organic inhibitor, namely, benzotriazole (BTA) was tested for the role of inhibitor in this study. Even at a BTA concentration as low as 0.005 M, the dissolution rate was reduced by three orders of magnitude. The very low dissolution rates obtained with the addition of BTA demonstrate that BTA is indeed a good candidate for the role of inhibitor in slurries based on H_2O_2 and glycine. The polish rate also decreased sharply with the addition of BTA and leveled off with further addition. Furthermore the *in situ* OCP measured with respect to SCE jumped towards a more anodic value when the polishing was stopped. This sharp increase in the OCP once the polishing was stopped indicates the formation of a passive film which could be either the oxide film formed because of the presence of H_2O_2 and glycine or a Cu-BTA inhibitor film. The fact that the dissolution rate is negligible in the presence of BTA indicates that it is indeed the Cu-BTA inhibitor film. Further more, the non-zero polish rate rates obtained with the addition of BTA indicates a dynamic interaction between the BTA and the *OH, with the BTA forming a protective film and *OH dissolving the copper. Thus, the addition of BTA protects the recessed regions where the mechanical abrasion will be minimal. Hence, in these regions the removal rate is comparable to the dissolution rate while the protruding regions are planarized by the combined action of abrasion by particles and removal by *OH leading to an efficient planarization process.

4. Summary

Copper dissolution/polish rates increased with increasing glycine concentration in the presence of hydrogen peroxide. The Cu dissolution/polish rates further increased with the addition of external copper ions confirming the role of Cu^{2+}-(glycine)$_2$ complex in catalysing peroxide to yield *OH radicals which plays a major role in Cu CMP. However, the copper dissolution/polish rates decreased with increasing peroxide concentration. This indicates the formation of a passive film during CMP which was confirmed by *in situ* electrochemical measurements. The effect of BTA as an inhibitor for copper dissolution during CMP using peroxide-glycine based slurries has also been demonstrated.

Acknowledgments

This research has been supported by a contract from NY State Energy Research Development Authority to Ferro Corporation and a grant from NY State Science and Technology Foundation to the Center for Advanced Materials Processing at Clarkson University

References:

1 Babu, S.V., Li, Y., Hariharaputhiran, M., Ramarajan, S., Zhang, J., Her, Y.S., and Prendergast, J.E., proc. 15th VLSI Multilevel Interconnection Conference (VMIC), June 1998, Santa Clara, CA

2 Prendergast, J.E., Mayton, M.M., Her, Y.S., Babu, S.V., Li, Y., and Hariharaputhiran, M., US patent (submitted on April, 1998)

3 Hariharaputhiran, M., Zhang, J., Li, Y., and Babu, S.V., submitted to Chemistry of Materials, feb. 1999

4 Hirabayashi, H., Higuchi, M., Kinoshita, M., Kaneko, H., Hagasaka, N., Mase, K., and Osima, proc. Of the 1st international VMIC specialty conference on CMP and planarization, Santa Clara, CA, 1996, p. 119

5 Hirabayashi, H., Higuchi, M., Kinoshita, M., Kaneko, H., Hagasaka, N., Mase, K., and Oshima, J., US Patent No. 5,575,885, Nov. 1996

6 Ramarajan, S., Hariharaputhiran, M., Her, Y.S., and Babu, S.V., proc. Surface Modification Technologies XII, ASM, 1998, p. 415-422

7 Luo, Q., Mackay, R.A., and Babu, S.V., Chemistry of Materials, 1997, 9, 10 p. 2101.

8 Srividya, C.V., Sunkara, M., and Babu, S.V., Journal of Materials Research, 1997, 12, 8, p. 2099, 1997

9 Stern, M., and Geary, A.L., J. Electrochem. Soc., 1957, 104, 1, p. 56-63.

10 Buxton, g.V., Greenstock, C.L., Helman, W.P., Ross, A.B., J. Phys. Chem. Ref. Data 17, 513, 1988

11 Kirk-Othmer, Encyclopedia of Chemical Technology, 4th edn., Vol. 5, p. 764-795, John Wiley & sons, New York, 1996.

12 Zeidler, D., Stavreva, Z., Plotner, M., and Drescher, K., Microelectronic Engineering, 33, p. 259-265, 1997

13 Luo, Q., Mackay, R.A., and Babu, S.V., Chemistry of Materials, 9, 10, p. 2101-2106, 1997.

SURFACTANT BASED ALUMINA SLURRIES FOR COPPER CMP

ASHOK K BABEL, RAYMOND A. MACKAY
Center for Advanced Materials Processing
Clarkson University, Potsdam, NY 13699

ABSTRACT:

The polishing of copper and examination of the polished surfaces were carried out with surfactant based alumina slurries to yield interesting results. Contrary to our expectation and previously reported research, some of the surfactant based alumina slurries resulted in higher copper polish rates when compared to the control. Of the nonionic surfactants, Brij[R] 35 was overall the most effective in both acidic and basic media. Ionics were effective at the pH for the appropriate charge type. For the range of surfactants studied, polish rates correlated with the HLB of the nonionic surfactants. The Hydrophile-Lipophile Balance (HLB) is related to the solubility of the surfactant, with higher number corresponding to increased water dispersibility. The surfactant Brij[R] 35, with the nonionic composition polyoxyethylene(23) lauryl ether, resulted in a dramatic improvement in the average surface uniformity when compared with the control at pH 2, and Sodium Dodecyl Sulfate produced even greater uniformity. Additionally, the effect of Brij[R] 35 surfactant was maintained with change in abrasive size, pad and polishing tool. In order to insure that surfactants are compatible with the chemical reagents contained in the commercial slurries, two chemistries (ferric nitrate and hydrogen peroxide) were employed to test the efficiency of the selected surfactants in their presence. The results showed that the effect of surfactant on stability and removal rate is not influenced by the presence of the chemicals. Preliminary results indicate that surfactants can have a beneficial effect on both defects and post polish clean.

INTRODUCTION

With the introduction of copper as an interconnect material of choice in the fabrication of microelectronic circuits, there is a considerable interest in the optimization of the planarization technique used to flatten the deposited copper thin films [3,8]. Chemical mechanical planarization (CMP) is identified as the most suitable method to achieve the global planarization. The CMP technique consists of rotating a polishing media against the rotating or stationary wafer while a polishing slurry is passed between the two surfaces. The pad is usually a stacked polymer structure which applied pressure to the wafer and carries the slurry between sample surface and the pad. A typical schematic of a CMP apparatus is shown in figure 1. The wafer is pressed against the pad with a known downward pressure. Polishing slurry is one of the main variables of the process affecting the polishing performance. The slurry consists of a dispersion of abrasive particles in deionized (DI) water with the added reagents (e. g. oxidizers, passivating agents, and other additives), adjusted to a desired pH. Addition of chemical reagents often causes the suspension to destabilize. Copper, being a soft material, is particularly vulnerable to scratches and defect often caused by presence of agglomerated abrasive particles. Physical modification of these abrasive particulates in order to enhance suspension stability, and thus to enhance uniformity across the polished wafer, and control the post polish surface defects (scratches and chemical and physical contamination) are the motivation behind this study. The use of surfactants for colloid suspension is well established in other areas [1]. Recently, there has been interest shown in the application of surfactants for CMP slurries as well [2,4-7]. This

Mat. Res. Soc. Symp. Proc. Vol. 566 ©2000 Materials Research Society paper describes the results we have obtained with the use of nonionic surfactants in alumina based slurries for the purpose of copper polishing.

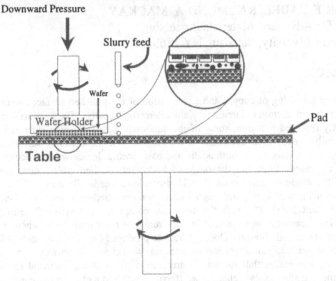

Figure 1 : Schematic of a chemical- mechanical polishing tool.

EXPERIMENTAL

Materials: The abrasive particles used in this study were obtained from Baikowski International Corporation and were used as received. The physical properties of these alumina abrasive particles are given in table 1. The particles size distribution of CR15 in the slurry was measured by means of a optic dynamic light scattering (DLS) using a Leed and Northrup Microtac UPA 150 and a mean diameter of 0.39 microns with a standard deviation of 0.15 micron was obtained. The surfactants were obtained from the Aldrich chemical Company. The names, chemical formula and other properties of these surfactants are outlined in table 2. The other chemical reagents used in the study were obtained from commercial venders (e. g. Fisher Scientific, J. T. Baker, Sigma Chemicals, etc.) and used as obtained. The slurries are prepared (2 wt % abrasives) with deionized water just before the experiments are carried out. The surfactant based slurries are kept over night for the surfactants to saturate the particles with bilayers.

Table 1: Physical Properties of the Bikalox Particles.

Product Code	CR6	CR15
Primary Particle Size (microns)	0.25	0.1
Unmilled Nominal Particle Size (micron)	1.0	0.3
Specific Surface Area (BET-m^2/gm)	6	15
Hardness (Mohs)	9	9
Crystal Density (gm/cc)	3.98	3.98
Bulk Density (gm/cc)	0.56	0.45
Major Crystal Phase, %	Alpha, 98	Alpha, 85
Crystal Structure	Rhombohedral	Rhombohedral

Table 2. List of Nonionic Surfactants

Product	Chemical Name	Chemical Formula[a]	HLB#[b]
Brij[R] 30	Polyoxyethylene(4) Lauryl Ether	$C_{12}E_4$	9.7
Brij[R] 35	Polyoxyethylene(23) Lauryl Ether	$C_{12}E_{23}$	16.9
Brij[R] 52	Polyoxyethylene(2) Cetyl Ether	$C_{16}E_2$	5.3
Brij[R] 58	Polyoxyethylene(20) Cetyl Ether	$C_{16}E_{20}$	15.7
Brij[R] 72	Polyoxyethylene(2) Stearyl Ether	$C_{18}E_2$	4.9
Brij[R] 92	Polyoxyethylene(2) Oleyl Ether	$C_{18}E_2$	4.9
Brij[R] 98	Polyoxyethylene(20) Oleyl Ether	$C_{18}E_{20}$	15.3
SDS	Sodium Dodecyl Sulfate	$C_{12}H_{25}SO_4Na$	40
CTAB	Cetyl Trimethyl Ammonium Bromide	$C_{16}H_{33}N(CH_3)_3Br$	

[a] C_n=Hydrocarbon chain length, E_m= Number of ethylene oxide units
[b] HLB= Hydropile-lipophile balance

2. Stability: The technique and results of stability experiments were described explained in an earlier paper [1]. In brief, a series of commercially available homologous nonionic surfactants of the polyoxyethylene alkyl ether type and two nonionic surfactants were examined for their ability to affect the CMP process with respect to enhancing slurry stability. The salient features of the stability study are incorporated in the following discussion.

3. Polishing: Table 3 outlines the polishing conditions and machine parameters employed to carry out the experiments of this study. The disk experiments were carried out on a Struer's DAP-V polisher using a 3 mm thick pure copper disk with a cross sectional area of 7.5 cm^2 . A table speed of 150 RPM was maintained but the disk was kept stationary. The slurry was stirred as it was being supplied to the pad. The pad was conditioned after each run for one minute with a 220 grit sand paper. The pad used was a Suba 500 obtained from Rodel, Inc. The experiments with wafers were carried out on a Strasbaugh 6 CA polisher. Sputter-deposited Copper films (3 micron) on 100 mm diameter single crystal Silicon wafers with a Ti barrier layer were commercially obtained (Silica-Source Technology, Corp. and Scientific Coating Labs) for this purpose. In this case also, the pad was supplied with stirred slurry and was conditioned, after each run, for a minute with sand paper. In both cases, the polish rate was obtained by weighing the polished disk/wafer before and after the experiments.

Table 3: Copper Polishing Conditions[1]

Polishing Tool	Struer's DAP-V	Strasbaugh 6CA
Polishing material	Cu Disk	Cu Wafer
Slurry Flow rate, ml/s	1	4
Pressure, N/m^2	4.3×10^4	2.7×10^4
Particle Load	2 wt%	2 wt%
Table/Quill Speed, RPM	150	40/40

[1] When present, the surfactant concentration is all cases was 0.01 M.

4. Surface Uniformity: A Sloan Dektak IIA profilometer and a Burleigh (Metris 2000) Atomic Force Microscope (AFM) was used to examine the polished copper surfaces of the wafers. The Dektek Profilometer is capable of a vertical resolution of 5 A^0. The stylus radius is 12.5 A^0 . The scanning rate was medium and a length of 250 micron was scanned in each run. The values of roughness reported here are an average of 10 such scans on each surface. The standard deviation of these values can be takes as an indicator of within surface nonuniformity. The Burleigh AFM has a vertical resolution of 10 A^0 with a 50 A^0 horizontal resolution. To rid the wafer of the slurry particles, which remain adhered to the surface even after deionized water cleaning, a brush cleaner was used prior to the roughness measurement with AFM. The roughness was averaged from the 3 values obtained from scanning at different locations on the wafer.

5. Effect of Chemical Addition: In order to insure that surfactants are compatible with the commercial slurries containing chemical reagents, two slurry chemistries (ferric nitrate and hydrogen peroxide) were employed to test the efficiency of the selected surfactants in the presence of chemicals. The results showed that the effect of surfactant on stability and removal rate is not influenced by the presence of the chemicals. Both hydrogen peroxide and ferric nitrate based chemistries were added to $Brij^R$ 35 based alumina slurries.

RESULTS AND DISCUSSION

1. Stability:

As indicated above, the effect of surfactants on slurry stability is reported elsewhere [1]. Summarized below are the essential features as well as the rationale for the surfactants selected for the further study.

For the nonionic surfactants, the maximum stabilization under either acidic (pH2) or basis (pH10) conditions was by a factor of 3 albeit with different surfactants. At pH2, $Brij^R$ 35 was the most effective, and was in the "middle of the pack" at pH10. Thus, from the point of view of using only one surfactant at all values of the pH, $Brij^R$ 35 is the best choice. As will be seen below, this remains true for the other (polish rate, surface roughness) criterion as well. The other nonionic surfactants employed were selected to span the entire HLB range in order to serve as a basis for comparison ($Brij^R$ 35 has one of the highest HLB). Only two ionic surfactants were studied and both provided a maximum stabilization by a factor of 5 at appropriate pH; anionic SDS at pH 2 and cationic CTAB at pH10. Note that the isoelectric point of the alumina is about 9. These ionic surfactants were also examined in this study.

2. Polishing:

Extensive polishing experiments of copper with the surfactant based alumina slurries have indicted that some of the surfactants lead to an enhanced removal rate. Others result in moderate to significant drops in removal rate. This trend seems to have some correlation with the slurry stability. Figure 2 shows the copper removal rate in nm/min as a function of HLB number for the surfactants used in this study at pH 2. Higher HLB numbers of nonionic surfactants generally was an indication of higher polish rate, but $Brij^R$ 52 is a notable exception. Other experiments have indicated that this enhancement (due to $Brij^R$ 35 addition) of removal rate is independent of pad type, abrasive, and polishing tool. Table 4 gives removal rates of copper with control (no surfactant) and with the $Brij^R$ 35 surfactant based slurry while a particular variable is changed. Table 4 shows that the copper removal rate is increased with two aluminas of different mean particle size. It also shows that effect of surfactant addition is independent of pad type (both Suba 500 and IC1000 show a increase in polish rate with $Brij^R$ 35 addition) and polishing tool (disk were polished on Struers polished and wafer were polished with Strasbaugh). The variation of effect of surfactant addition with a change in media pH (by addition of acid, base or buffer) on polishing has also been determined. For example, $Brij^R$ 35 increases the removal rate in acidic media but it maintains the same polish rate in neutral and basic slurries compared to the non-surfactant systems. This is true irrespective of how was pH was adjusted (either acid/base or buffers). Figure 3 shows the removal rate of copper with slurries of CR15 made a different pH with buffers.

The surfactants that lead to destabilization of the slurry also lead to a lowering of the removal rate but Brij^R 30 and particularly Brij^R 52 are notable exceptions. However, as discussed below, But the use of Brij^R 52 and Brij^R 30 based slurries also results in a surface with higher average nonuniformity. In contrast, the high polish rate of Brij^R 35 based slurry is also accompanied by a high uniformity of the polished surface as compared to nonsurfactant slurries.

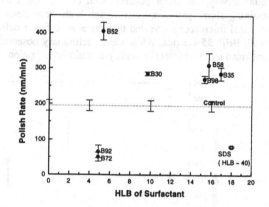

Figure 2: Removal rate as a function of surfactant HLB. The data are for the copper disk, polished on Struers (Table 2) at pH 2. The ionic surfactant SDS is included, although off the 0-20 nonionic HLB scale.

Table 4: Effect of Tool, Abrasive and Pad on Surfactant effect (Removal rate in nm/min, pH=2)

	Polish Tool[1]		Abrasive[2]		Pad[3]	
	Struers	Strasbaugh	CR15	CR6	Suba 500	IC 1000
No Surfactant	195±15	148±10	195± 15	198± 12	195± 15	120± 5
With Brij^R 35	285± 18	291±18	285± 18	462± 30	285± 18	240± 27

[1]CR15+Suba Rodel Pad
[2]Struers+Suba Rodel Pad
[3]Struers+CR15

3. Surface Uniformity:

The results indicate that some surfactants can definitely lead to better surface finish upon polish. Table 5 gives the average RMS (root mean square) values, Ra, of surface roughness obtained with the Dektak profilometer. The measurements are carried out on both the copper disks and 4 inch copper wafers. Dramatic improvements are noticed in polished surface uniformity when surfactants Brij^R 35 and SDS are employed in the slurry. The destabilizing surfactants Brij[R] 72 and Brij[R] 52 (Brij[R] 52 giving a high polish rate) produce twice as nonuniform a surface as compared to the surfaces polished with non surfactant slurry. The degree of reduction in nonuniformity (standard deviation of roughness) for both copper disks and copper wafers was greater then 50%. Notice the rows with plain and Brij[R] 35 added slurry in both disk and wafer case in table 5. This result is also obvious from figures 4a and 4b which show the sample AFM images of polished copper wafer surfaces. Figure 4a is one which is polished with control slurry (no surfactant added). Figure 4b shows a similar scan of the wafer polished with Brij[R] 35 surfactant based slurry at buffer pH 2. Although there are still particles remaining on the wafer, the surface in the figure 4a is clearly smoother then in figure 4b. It is also noted that, qualitatively, the wafer polished with the surfactant based slurry appears to contain fewer remaining slurry particles, even though both were cleaned with the same post clean procedure. Optical microscope revealed that there were fewer defects (scratches) on the surface polished with Brij[R] 35 slurries. While these preliminary observations are encouraging, they need to be confirmed by more intensive and systematic investigation.

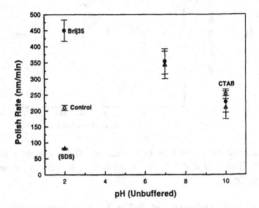

Figure 3: Removal rate of Brij[R] 35 based slurries as a function of slurry pH. The ionic surfactant SDS and CTAB is included, although off the 0-20 nonionic HLB scale.

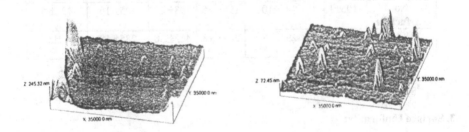

Figure 4a and 4b: AFM scans of a copper wafer polished with plain and surfactant based alumina slurry (average roughness = 9 nm for 4a and =4 nm for 4b).

Table 5: Surface Roughness of Copper Surfaces Polished with Surfactant based slurries (Scan length=250 micron)

	Slurry	Roughness Ra, nm
Disk Polish (Streur's)	Before Polish	165± 82
	Plain Slurry	45± 23
	Brij[R] 35	19± 2
	Brij[R] 52	85 ±29
	Brij[R] 30	75±27
	Brij[R] 72	88±28
	SDS (pH=2)	9±2
	CTAB (pH=10)	35±9
Wafer Polish (Strasbaugh)	Before Polish	3.3±0.1
	Plain Slurry	16.6±5.6
	With Brij[R] 35	6.5±2.0

4. Effect of Chemical Addition:

The stability of the slurries was unaffected (as compared to the non chemical slurries), whereas the enhancement of removal rate attributed to the Brij[R] 35 is maintained. Table 5 gives the polish rate of the copper when chemicals are added to the slurries. In the case of hydrogen peroxide, the chemical contribution to removal rate is maintained (from 835 nm/min to 850) with the addition of Brij[R] 35 (this was the case with nonchemical slurries, from 343 to 354). But for the case of ferric nitrate (at pH=2), the increase of the polish rate is also recaptured with addition of Brij[R] 35 (from 310 to 670 as compared to 219 to 450). These results are consistent with the effect of pH alone on the removal rate.

Table 6: Change in surfactant effect on chemical addition

Chemical Basis	Slurry	Polish Rate, nm/min
Hydrogen Peroxide[1] (pH=6)	CR15	343±42
	CR15+Brij[R] 35	354±39
	CR15 +Chemicals	835±5
	CR15+Chemical+Brij[R] 35	850±15
Ferric Nitrate[2] (Ph=2)	CR15	219±36
	CR15+Brij[R] 35	450±33
	CR15+ Chemicals	310±15
	CR15+ Chemical +Brij[R] 35	670±15

[2]Ferric Nitrate: 0.1M, Benzo-triazole (BTA): 0.005M
[1]Hydrogen Peroxide: 5 wt%, Glycene: 1 wt%, BTA: 0.005M

CONCLUSIONS

Surfactants which stabilize the alumina suspensions either maintains or increases the copper removal rate when polished with these slurries, as compared to the corresponding surfactant-free slurries. This is maintained in the presence of added chemistries (hydrogen peroxide or ferric nitrate), and is independent of the abrasive surface area, pad, and polishing tool. Surfactants which stabilize the slurry also results improved surface uniformity after polishing. Some surfactants which destabilize the slurry also yield increased removal rate but a significantly degraded surface uniformity. This may be due to aggregated particles which does not readily break up under polishing conditions. Very preliminary observations indicate that surfactants can also reduce defects and enhance post-polish clean. Of the nonionic surfactants examined, Brij^R 35 was the optimal choice at all pH. The ionic surfactants, when their use is compatible with other requirements, can also be useful at appropriate pH.

Future work will examine the effect of surfactants on defects and post clean process, as well as the mechanism(s) of action.

ACKNOWLEDGMENTS

The authors wish to thank the Center for Advanced Materials Processing (CAMP), at Clarkson University and to the SRC via the CAIST at RPI, for the support of this work. They also thank Prof. S. V. Babu for the use of his polishing and profiling instruments.

REFERENCES

1. Babel, A. K., Campbell, D. R., Barry, J. D., and Mackay, R. A., submitted for publication.
2. Bielmann, M., Mahajan, U., Singh, R. K., Shah, D. O., and Palla, B. J., *Electrochemical and Solid-state Letters*, 2(3), 148, 1999.
3. A.E. Braun, *Semiconductor International*, **21**, 65-74, 1998.
4. G. Sabde, *Private Communication*.
5. M.L. Free, D.O. Shah, *Micro*, 29-37, May 1998.
6. B.J. Palla, D. O. Shah, M. Bielmann, and R.K. Singh, *Private Communication*.
7. J. J. Adler, Y.I. Rabinovich, R. K. Singh, and B.M. Moudgil, *Mat. Res. Soc. Symp. Proc.*, **501**, 387, 1998.
8. D. R. Campbell, K. Achutan, and S. V. Babu, *CAMP Newsletter*, **10**, 1, 1994.

IMPACT OF LOW-TEMPERATURE ANNEALS OF ELECTROPLATED COPPER FILMS ON COPPER CMP REMOVAL RATES

KONSTANTIN SMEKALIN AND QING-TANG JIANG, Sematech Inc., 2706 Montoplis Dr., Austin, TX 78741, and National Semiconductor Inc., 2900 Semiconductor Dr., Santa Clara, CA 95052.

ABSTRACT

CMP removal rate (RR) of electrodeposited Cu film was found to increase by 35% over time after plating. The RR increase was attributed to Cu film hardness reduction of 43% and grain growth from the initial 0.1um as-deposit to 1um at the final stage at room temperature. The removal rate increase will translate to variations in manufacturing environment and are therefore unacceptable. It was found that annealing at ~100C for 5 minutes in inert gas will stabilize Cu films and provide consistent CMP removal rate.

INTRODUCTION

Copper is rapidly advancing to replace Al alloys as interconnect material in IC manufacturing due to the advantages of Cu in electrical resistance and electromigration. At the same time, implementation of Cu also promises to reduce the cost of the back-end wafer processing by utilizing damascene technology. In most practical applications, electroplating of Cu is used to fill the trenches, following liner and seed layer deposition. Interconnect wires are then delineated by Chemical Mechanical Polishing (CMP) of Copper. Electroplating of Cu and Cu CMP play key roles in developing reliable interconnects. Material properties of electroplated Copper and their impact on Copper CMP therefore become critical for successful utilization of Copper damascene technology.

It has been observed earlier [1,2], that electroplated Cu films undergo self-induced changes at room temperature, which results in a 20% reduction in bulk resistivity and grain growth from ~0.1 μm to ~1.0 μm. It is critical to understand that in a production environment this change will impact manufacturing. In this paper, we discuss how the self-induced change in mechanical properties of electrodeposited Copper films impact Cu CMP removal rates (RR). We suggest annealing at temperatures around 100°C for several minutes as a means to stabilize Copper films, and provide consistent RR for the subsequent CMP.

EXPERIMENT AND RESULTS

Cu electroplating was performed using sulfate based plating bath with a small amount of organic additives. Prior to Cu plating, a thin metal barrier layer and 0.1 μm Copper seed layer were deposited on 200 mm Silicon wafers with 0.8 μm buffer oxide layer. The thickness of the electrodeposited Cu was measured using cross-sectional SEM to be 1.6 μm. To exclude random variations in CMP process, before each test wafer was processed, the removal rate was calibrated using fully stabilized blanket Cu wafers. The amount of Cu removed from the wafer by CMP process was determined by measuring the weight loss by analytical balance, or measuring remaining

As observed earlier [2,3], electrodeposited Copper films undergo self-induced transformation after electroplating which results in resistivity reduction and grain growth. Mechanical properties of the films also change. Using microhardness tester with a diamond pyramid Vickers indenter, we measured the film hardness change as a function of storage time at room temperature.

143

Hardness measurements were taken at vartious time intervals. During the measurement, a known load was applied and a permanent indentation was created on the material under measurement. The hardness value was extracted by measuring the size of the indentation using an optical system. Since our Copper films were thin and soft, the applied deformation load was small. The applied load ramped from 0.04 to 0.4 grams using 15 load and unload steps with a time interval of 4 seconds between steps.

Fig.1 displays these hardness measurements as a function of storage time. For reference, the resistance transformation is displayed on the same graph (dashed line). It was observed that the film hardness dropped 43% over time of 50-60 hours. It started with an initial hardness reading of 280 Hv and dropped to 160 Hv* in the same time frame of resistivity transformation. It is necessary to find out how much this hardness reduction will impact the CMP removal rate since it has been reported that film hardness can have a profound influence on CMP removal rate [4].

Hardness Change of 1.6μm EP Cu Film

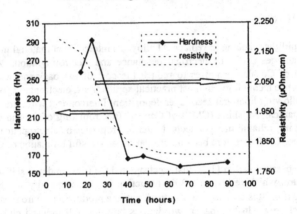

Fig. 1 Cu film hardness change as a function of time for a 1.6μm electrodeposited Cu film at room temperature.

To investigate CMP removal rate change as Cu films self anneal, a batch of blanket wafers was electroplated under the same condition. As Cu films were being aged, wafers were pulled at different times to go through an identical CMP process. Cu removal for each wafer was extracted by measuring the initial and final thickness using cross-sectional SEM and also independently measuring wafer weight loss with a balance. In order to identity the stage of transformation, a wafer from the same batch was pulled and monitored regularly for sheet resistance change through the entire period.

Fig.2 displays the normalized Cu removal rate as a function of storage time. For reference, the sheet resistance self-annealing curve is also displayed in the graph. As the graph demonstrates, during film transformation, the CMP removal rate increased by ~35%. The increase of Cu removal rate over time was in part caused by the 43% mechanical softening of the

---

* Vickers Hardness Hv is defined as the ratio of the applied load to the indented "unrecovered" projected area. The higher the Hv value, the harder the film is.

Cu film. It is obvious that room temperature self-annealing would result in high variations in CMP process in a real production environment, which is unacceptable for IC manufacturing. In order to ensure consistent Cu film properties and stable CMP removal rates, the electroplated Cu films have to be stabilized.

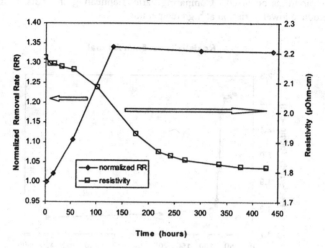

Fig. 2 CMP removal rate increased 35% during Cu film self annealing.

Using a Blue M inert gas oven, we investigated acceleration procedures to bring Cu films to a quick stabilization. Given the situation that Cu films evolve at room temperature, it is assumed that the temperatures required to stabilize the film will not need to be high. It is important to realize that Cu readily reacts with oxygen in air. The oxidation gets worse even at slightly elevated temperatures. Therefore, the anneal has to be done in a controlled environment. Wafers from the same plating batch were annealed for 5 minutes at different temperatures in inert gas. Sheet resistance was measured using a 4-point probe and converted to resistivity after measuring the film thickness using SEM. The results are displayed in Fig.3. It showed that the bulk resistivity of an as-deposited film was around 2.2 μOhm·cm. It dropped rapidly to 1.8 μOhm.cm in the temperature range from 55°C to 70°C. Further anneals at higher temperatures, up to 450°C, only slightly reduced the resistivity further to 1.75 μOhm.cm, which was fairly close to the theoretical Cu resistivity of 1.7 μOhm.cm. Fig.4 demonstrats the stress measurement from the same wafers. Our plating process consistently produced films with small compressive stress on the order of $1.0 \cdot 10^8$ dynes/cm2. The stress became tensile as annealing temperature rose. The film finally reached a tensile stress of $2.0 \cdot 10^9$ dynes/cm^2 at ~100°C, which was still considerably small compared to PVD Copper films [2]. Further increase of temperature did not change the film stress anymore.

Fig. 5 shows the normalized CMP removal rate of electroplated 1.6 μm Cu films after 5 minutes anneal at different temperatures in both N2 and forming gas (95% N2 and 5% H2). A 30 minutes inert gas purge was employed prior to temperature ramp for anneals at temperatures higher than 120°C. The amount of Cu removed from the wafer was extracted by weighing the wafer before and after the CMP process. Fig.5 indicated the CMP removal rate increased rapidly as temperature increased and the Cu film went through different transformation stages. The insert showed the details of removal rate in the temperature range from 70°C to 110°C. It showed a relatively stable region in this temperature range. However, the removal rate did not remain stable at higher temperatures. It became unstable at temperatures higher than 120°C, where large variations occurred. Comparing various annealing ambients, forming gas anneal seemed to produce lower variation at high temperatures.

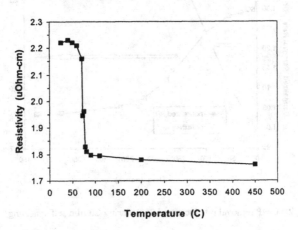

Resistivity vs. 5min Anneal

Fig. 3   Bulk resistivity dropped rapidly when annealed 5 minutes at increased temperatures.

Since the inert gas oven used in the experiments was an atmospheric pressure oven, the door seals were not totally leak-tight. In addition, the oven operate only at 0.5 psi higher than the atmospheric pressure. Even with trace amounts of oxygen, Cu oxidation is definitely a concern. It was often noticed that Cu changed color from bright to dull after anneal. Using cathodic stripping method, we measured the amount of copper oxide growth on the wafer after each annealing process. It is worth mentioning that the oxide measurements were performed more than 30 days after annealing experiment was done due to the unavailability of the oxide measurement. The measurement showed only Cu_2O presented. The amount of CuO was negligible. The thickness of Cu_2O after anneal at different temperatures in both N2 and forming gas ambient was found below 100 Å. Also, the slightly lower thickness of Cu_2O in forming gas at temperatures higher than 120°C was because of the purge step employed for temperatures higher than 120°C.

Stress vs. 5min Anneal

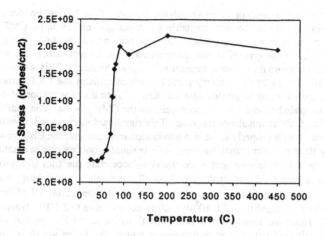

Fig. 4   Film stress changed from small compressive to tensile during anneal.

Anneal Effect on CMP

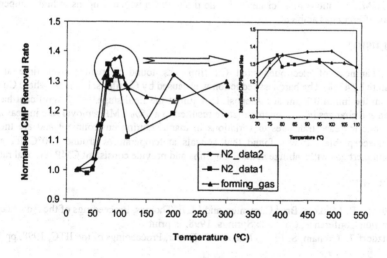

Fig. 5   CMP removal rate change as a function of 5 minutes anneal at different temperatures.

SUMMARY

The process of self-annealing of electroplated Copper films has to be dealt with in a production environment as it produces unstable Cu films and causes large CMP variations. Reduction in film hardness, which amounts to 43%, appears to be a direct result of grain growth from 0.1um to 1um since the grain size increase makes it easier for dislocations in the film to slip over longer distances before they have to interact with a grain boundary. This softening promotes microindenting of the film by CMP slurry particles, which initiates the CMP removal. At the same time, grain growth in self-annealed film can also enhance the propagation of the chemical reactions along the grain boundaries and thus accelerate the CMP process. The combined effects resulted in the 35% CMP removal rate increase. This significant increase will translate to CMP variations if not dealt with properly since in a production environment, the CMP process can be performed at any stage of a film transformation. This is simply unacceptable in manufacturing.

Anneals around 100°C in inert gas were found to accelerate the film transformation and stabilization in minutes for 1.6 μm blanket Cu films. The anneals will also bring other physical parameters to a final or near final values, where resistivity dropped from 2.2 μOhm.cm to 1.8 μOhm.cm, film stress changed from $1.0 \cdot 10^8$ dynes/cm^2 compressive to $2.0 \cdot 10^9$ dynes/cm^2 tensile. The significantly increased tensile stress after anneal will need to be checked out for void formation and/or side wall detachment in damascene structures. Large variations of anneals at temperatures higher than 120C are possibly due to oxidation which happens more easily at these temperatures in different gas ambient. It is also possibly due to the microstructural difference after annealing at different temperatures. Forming gas anneal seemed to work better in terms of process stability at higher temperatures than N2 anneal. Although figure 6 did show differences in oxide growth for anneals at different temperatures in different ambient, the differences are not the root cause of CMP variations but instead, indications of differences in the bulk Cu. The root cause of CMP variations must originate inside the bulk Cu which happens at high temperatures and/or in different gas ambient.

CONCLUSION

Film hardness of electrodeposited Cu film was found to reduce over time at room temperature by 43%. The hardness reduction was caused by Cu film self anneal where Cu grains grow from the initial 0.1 μm at as-deposit to 1 μm at the final stage. The significant hardness reduction and changes in film microstructure resulted in a 35% CMP removal rate increase. This removal rate increase translates to variations in manufacturing environment and are therefore simply unacceptable. It was found that anneals at temperatures around 100°C for several minutes in inert gas will stabilize blanket Cu films and provide consistent CMP removal rate.

REFERENCES

1. Q.T. Jiang, R. Mikkola, Brad Carpenter, and M. E. Thomas, Proceedings of the Advanced Metallization Conference, Colorado Springs, 1998, in print
2. T. Ritzdorf, L. Graham, S. Jin, C. Mu, and D. Fraser, Proceedings of the IITC, 1998, pp.166-168.
3. Frank G. Shi, Bin Zhao and Shi-Qing Wang, Proceedings of the IITC, pp. 73-75.
4. Paul A. Flinn, J. Mater. Res. Vol. 6, No. 7, Jul 1991, pp. 1498-1501.

Modified Preston Equation- Revisited

S.Ramarajan[*] and S.V.Babu[*]
[*]Center for Advanced Materials Processing, Clarkson University, Potsdam, NY-13699.

ABSTRACT

The effect of pressure and velocity on the polish rates of copper was determined in DI water and in the presence of ferric nitrate, H_2O_2/glycine, and NH_4OH with alumina particles as the abrasives. The polish rate shows a stronger dependence on velocity than that predicted by the Preston equation in the case of ferric nitrate, a highly reactive chemical. The velocity dependence is weaker for the other two less reactive chemicals, and is the same as that predicted by Preston equation for DI water. Our earlier empirical model, $R = KPV + BV + R_c$, where K, B, and R_C are constants, describes all the polish rate data satisfactorily.

INTRODUCTION

The effect of pressure and velocity on the removal rate of oxide and metal films in a chemical-mechanical polish process has been widely investigated. Yet, there is no consensus and several different theoretical models have been proposed. Most of them take the form $R = KP^aV^b$ [1-2], where R denotes the polish rate, K is a proportionality constant, P is the applied downward pressure, V is the velocity, and a and b are constants that take fractional values between 1/3-1. When a and b are both equal to 1, this equation becomes the Preston equation[3], well known in optical glass polishing industry.

Liu et. al.,[4] modeled the wear mechanism during chemical-mechanical polishing (CMP) and reported the Preston equation to be valid for polishing of SiO_2 films. However, Zhao and Shi[5] reviewed the polish rate dependence on pressure and velocity and argued that the relative hardness of the pad with respect to the wafer and the abrasive particles plays a crucial role. It is suggested that Preston's equation, with its linear dependence on pressure and velocity, is applicable only when the hardness of the polishing pad is similar to or higher than that of the abrasives or the surface being polished. They show how the rate dependence on pressure can become sub-linear if the pads are softer and derive a value of 2/3 for the exponent a based on the contact area between an asperity and the wafer surface. Furthermore, according to them, while polishing with such soft pads, the abrasive particle can slide and cause appreciable material removal only when the applied pressure exceeds a threshold value. Since in a typical CMP process in silicon device manufacturing, the pads are always softer than the films being planarized, they argued that Preston equation needs to be modified for describing the polish rate dependence on pressure. Indeed, they were able to fit the pressure dependence of the measured polish rates for SiO_2, Cu and Al reported by several authors using a non-zero threshold pressure and a value of a = 2/3.

In all these derivations, the role of the slurry chemicals during the polish process is not apparent. Even under static conditions, some of the chemicals can dissolve the material as in the case of ferric nitrate and copper or even H_2O_2/glycine and copper. This effect can, in principle, be easily included in a model description by adding a nonzero, velocity and pressure independent, intercept to the polish rate expression. In practice, it is more complicated since the relation between this nonzero intercept and static dissolution rates is not simple and is unknown due to, among other things, the effects of the polishing pad. In such cases, the role of a threshold pressure, while perhaps obvious when mechanical abrasion is the only mechanism for material removal, is not evident unless the removal rate can be broken neatly into two *independent* terms, one for the mechanical abrasion and the second for the chemical removal. Such is the case for the

Mat. Res. Soc. Symp. Proc. Vol. 566 ©2000 Materials Research Society polishing of copper in ferric nitrate and alumina slurries, where we showed earlier that the mechanical and chemical removal rates are additive and there is no synergy. In such cases, the threshold pressure term would be included in the expression for the mechanical removal rate.

However, mechanical and chemical removal rates are not always additive, as in the case of polishing of copper in peroxide or ammonia based slurries. These chemicals induce changes in the film surface composition and hardness. In other words, the chemical and mechanical removal mechanisms are coupled and, if so, the above method of modifying the Preston equation may not be appropriate. Additionally, as argued below, it is, a priori, reasonable to expect that the polish rate may have a stronger dependence on velocity than that indicated by the Preston equation, especially when the chemicals in the slurry play a significant role in the removal process. We present new polish rate data for copper that supports this stronger velocity dependence and show that an empirical expression proposed by us earlier describes this data satisfactorily. We conclude with a discussion of some possible implications for the polishing of other materials.

EXPERIMENTS

Copper films for the polishing experiments were sputter deposited using a TORR International CRC-150 sputtering system on 6" blanket silicon wafers with tantalum as the adhesion promotion layer. The resistivity of the sputtered copper films was determined to be 2.5 $\mu\Omega$ cm. The slurries for the polishing experiment were prepared using α-alumina particles, with a bulk density of 3.7 g/cc, obtained from Ferro Corporation. The solids concentration was kept at 2 wt % unless otherwise stated. All the chemicals were purchased from Aldrich chemical company and were used without further purification.

The polishing experiments were carried out using a Strasbaugh 6DS-SP polisher at Ferro and a Strasbaugh 6 CA polisher at Clarkson University. The polisher parameters were set at no oscillation and 240ml/min slurry flow rate for all the polishing experiments. The downward pressure was varied between 2 to 9 psi and the rotational speed of the table and quill between 10 to 70 rpm. In each experiment the rotational speed of the table and quill remained the same. The polishing experiments were performed with a SUBA 500 polishing pad. The pad was conditioned using a diamond conditioner. The conditioner parameters on the 6DS-SP were set at 1.5 psi downward pressure, 4 sweeps and 20 seconds per sweep.

The thickness of the copper films before and after polishing was measured with a Signatone semi automatic four-point probe at 45 points on the wafer. The thickness reported was the arithmetic average of those 45 points and each reported rate was obtained by averaging the results from three runs at the same polishing conditions.

RESULTS AND DISCUSSION

Velocity Dependence

During polishing, especially of metallic films, the concentration of the slurry chemicals in the liquid layers between the pad and the wafer is an important variable in determining the material removal rate. Unlike the hard abrasive powders in the slurry, the chemicals are consumed and need to be constantly replenished. The velocity of the wafer and the pad will influence the concentration as well as the rate of replenishment of the chemicals in the liquid film entrained between the pad and the wafer[6]. Furthermore, the temperature of the film surface being polished[7] will very likely increase as the wafer/pad velocity is increased, which, in turn, will increase the rates of the chemical reactions occurring at the film surface. Hence, one may anticipate that the removal rate will have a much stronger dependence on the velocity than that predicted by the Preston equation, especially if the slurry chemicals play a dominant role in material removal from the film surface.

We have recently proposed an empirical modification[8], $R=KPV + BV + R_C$, of the linear Preston equation where B and R_c are constants, to describe some of our copper polish rate data. This expression was used to fit the polish rates of copper obtained using a ferric nitrate slurry and a H_2O_2-based commercial slurry (QCTT1010). A detailed discussion on the effect of slurry chemicals on the non-zero intercept was also given. The non-zero intercept changes from a high value for the case of dissolving chemicals all the way to zero or negligible value for the case of passivating chemicals. As suggested earlier, the higher velocity dependence could have resulted from the faster replenishment of slurry chemicals between the pad and the wafer. This implies that the relative magnitude of B (coefficient of the extra velocity term) for copper polishing will be higher for strongly dissolving chemicals like ferric nitrate than for H_2O_2/glycine or DI water. As the value of B changes, the relative importance of this term in relation to the Preston KPV term also changes. The results described below indeed confirm this for copper polishing. A similar relationship was also propose by Wu and Cook[9], who suggest that the copper removal rate consists of three additive components, namely, mechanical, hydrodynamic, and chemical, given by k_1PV, k_2V, and k_3, respectively.

Figure 1 and 2 show the polish rate of copper in the absence of chemicals as a function of velocity and pressure, respectively. The experiments were carried at two different pressures as a function of velocity and two different velocities as a function of pressure. The reason for the saturation of polish rates at higher velocity is not clear and these data points were not included in the calculation of slope of the velocity plot. The nearly equal slopes obtained from the two figures means that the rate dependence on velocity is not stronger. This combined with the near zero intercept obtained at zero pressure and velocity validates the Preston equation in this case of polishing in DI water. However, because of polish rate saturation at higher velocity, the Preston equation is valid only in a limited velocity range.

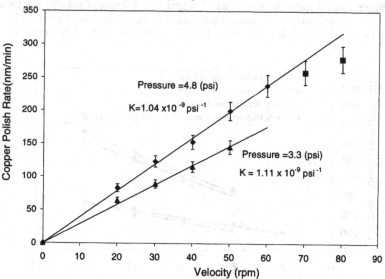

Figure 1: Copper polish rate as a function of velocity in de-ionized water

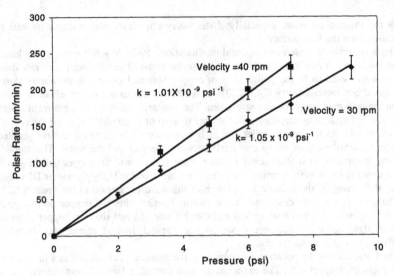

Figure 2: Copper polish rate as a function of pressure in de-ionized water

Polishing of copper was also carried out in the presence of three different solutions, 0.02M ferric nitrate, 5 wt % H_2O_2 and 0.2 wt % glycine, and 3 vol % of NH_4OH, all in DI water. These three solutions cover a wide dissolution regime, varying from a high dissolution rate in the presence of ferric nitrate to very low dissolution rate in the presence of ammonium hydroxide.

Figure 3 and 4 show the polish rate of copper as a function of velocity and pressure, respectively, in the presence of the above chemicals. The experiments were conducted at two different velocities and two different pressures for 0.02 M ferric nitrate but only one velocity and

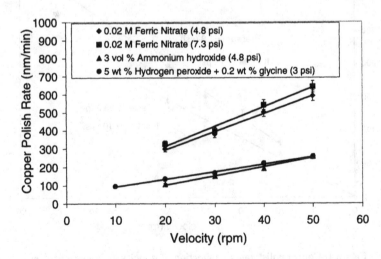

Figure 3. Copper polish rate as a function of velocity in the presence of various chemicals

one pressure for the other two solutions. The empirical expression $R = KPV + BV + R_C$ was fitted to the experimental data and the three parameters K, B and R_C were determined by least squares procedure. From figure 3, the values of the slope (KP + B) and intercept (R_C) were obtained for each value of P. From figure 4, the values of the slope (KV) and intercept (BV + R_C) were obtained for each velocity. The parameters R_C and B were then calculated from the two intercepts. Then using the value of B, the coefficient K was determined from the slope of figure 3 and compared with that determined independently from the slope of figure 4. The values of all the constants are listed in Table 1. Since, in the case of ferric nitrate, rates were measured at two different velocities and pressures, this procedure can be carried out with each set of pressure and velocity values and the resulting constants are also reported in table 1. The fact that the independent determination of the three constants yield values that are within ± 10 % of each other strongly suggests that this empirical equation provides a consistent representation of the above data sets.

Table 1. The parameters of the modified Preston equation.

Slurry composition			K Psi^{-1}	B	R_c (nm/m in)
0.02M Ferric nitrate	Constant pressure plot	4.8 psi	6.2×10^{-10}		94.2
		7.3 psi	5.7×10^{-10}		95.5
	Constant velocity plot	30 rpm	5.9×10^{-10}	9.5×10^{-9}	
		40rpm	5.4×10^{-10}	8.9×10^{-9}	
5 wt % H_2O_2 + 0.2 wt % Glycine	Constant pressure plot	3 psi	1.3×10^{-9}		40.6
	Constant velocity plot	30 rpm	1.3×10^{-9}	5.6×10^{-10}	
NH_4OH	Constant pressure plot	4.8 psi	1.3×10^{-9}	----	----
	Constant velocity plot	30 rpm	1.3×10^{-9}	1.4×10^{-10}	-----

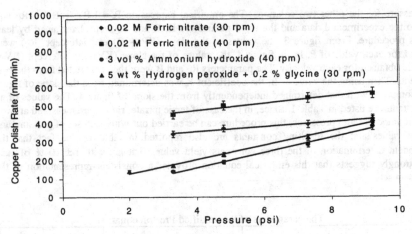

Figure 4: Copper polish rate as a function of pressure in the presence of various chemicals

CONCLUSIONS

The polish rate has a stronger dependence on velocity than that suggested by Preston equation, when the slurry chemicals play a significant role in material removal during chemical-mechanical polishing as in the case of metal films. The empirical removal rate, R=KPV + BV + Rc, where K, B and Rc are constants provides a satisfactory description of all the polish rate data presented here.

ACKNOWLEDGEMENTS

This research was supported by a contract from NY State Energy Research and Development Authority. The authors would like to thank the Ferro Corporation for allowing us to use their facility to conduct some of the experiments reported in this work.

REFERENCES
1. Wei-Tsu Tseng ., and Ying-Lang Wang, Journal of Electrochemical Society, 1997, **144**, p. L15.
2. Zhang, F., and Busnaina,A., Electrochemical and Solid-State Letters, 1998, p.184.
3. Preston, F., Journal of Society of Glass Technology, 1927, **11**, p.247.
4. Liu, C.W., Dai, B.T., Tseng, W.T., and Yeh, C.F, Journal of Electrochemical Society, 1996, **143**, p.716
5. Zhao, B., and Shi, F.G., Proc. 4th international CMP-MIC, p13-22, Santa Clara, CA, Feb, 1999.
6. Subramanian, R.S., Zhang, L., and Babu, S.V, Submitted to Journal of Electrochemical Society.
7. Stein, D.J., Hetherington, D.L., and Cecchi, J.L, Journal of Electrochemical Society, 1999, 146, p. 376.
8. Luo, Q., Ramarajan, S., and Babu, S.V, Thin Solid Films, 335, 1998, p.160
9. Wu, G., and Cook.L, Proc. 3rd international CMP-MIC, p.150, Santa Clara, CA, Feb, 1998.

AN EXPLORATION OF THE COPPER CMP REMOVAL MECHANISM

PETER RENTELN, TON NINH
Advanced Products Research and Development Laboratory (APRDL),
Motorola, 3501 Ed Bluestein Blvd., Austin, Texas 78721

ABSTRACT

Copper CMP is emerging as the next generation process technology enabling feature size reduction to .15μm and beyond[1]. We propose a copper removal mechanism in the context of a slurry consisting of an oxidizer and an abrasive. The body of evidence suggests that we are polishing in an oxidation complex rate limited regime. We observed low removal rate of copper in the absence of either oxidizer or abrasive, but rate was still dependent on CMP parameters and strongly tied to temperature. Any proposed mechanism must explain the observed dependence of rate on CMP aggressiveness and the role of each of the components. For the slurry used in this work we propose that an increase in temperature resulting from an increase in CMP intensity drives the kinetics of the oxidation reaction, and that the removal process can be classified as temperature-activated, abrasion assisted dissolution.

INTRODUCTION

There is an increasing body of literature on the copper CMP process[2-8], approaching the issue both from a phenomenological and a fundamental point of view. To be sustained, any model in which a removal mechanism is advanced must be either supported by all - or at least not contradicted by any - available observations. These observations generally include CMP rate and nonplanarity (dishing) effects as a function of CMP parameters such as speed and pressure, or as a function of slurry components such as oxidizers, abrasives and etchants. One model[9,10], for example, offers that the observation of low removal rate when polishing with abrasive only is due to the effect of redeposition of copper, which reduces the net removal rate. In this work we advance a model and evidence for it which explains removal as a function of oxidation of the copper surface and the corresponding creation of a soluble solid oxidation complex. While the model does not expressly disallow redeposition to occur, it is not required to explain low removal rate. The essence of the model is that friction caused by the interaction of the wafer and the pad heat the slurry to an elevated temperature without the addition of any intentional introduction of heat (the role of friction in oxide CMP has been investigated, e.g. ref. 11). And that while room temperature slurry may show no etch rate, when activated by temperature the slurry becomes an aggressive oxidizer/etcher (an investigation carried out simultaneously to this work reveals similar effects [12]). It is under these conditions that copper CMP proceeds at acceptable removal rates. This phenomenon also may affect dishing, although that topic is outside the scope of this work.

EXPERIMENTAL

Temperature and Rate

Unpatterned electroplated copper wafers were polished using an IPEC 472 polisher equipped with an IR temperature sensor which measures pad surface temperature. The copper rate was determined from copper thickness measurements from a Tencor RS55 resistance monitor, calibrated to cross-sectional SEM micrographs. The SEM tools are regularly calibrated to national standards.

A sequence of wafers was polished for incrementally increasing polish times. Measurement of thickness before and after polishing allowed for a determination of the integrated rate as a function of time. The integrated rate was observed to start low and increase rapidly, leveling off after about 40 seconds (diamonds in Fig. 1, below). To ensure that this was a CMP-related phenomenon and not intrinsic to the film, the experiment was repeated on wafers which had been first polished for 20 seconds (triangles in Fig. 1, below). It can be seen that the curves are fundamentally the same shape. Since the behavior repeats at a lower position in the copper film, it is clearly due to the CMP process and is not a film property.

Plotted with the rate is pad surface temperature (x's in Fig. 1), as measured for similar process conditions. The stark similarity of the shape of the curves adduces a strong hint as to the nature of the removal rate. Since the process conditions are not changing during this ramp, a direct causal temperature/rate relationship can clearly be inferred.

As a matter of interest, wafer uniformity is also plotted. The unit of measure is the slope, which derives from a manipulation of the standard 13 point measurement. A center point is measured, followed by three sets of four points, each set at a single radius. A regression consisting of the center point and the averages of the subsequent three "rings" of four points provides a measure of the concentric nonuniformity. Note that the wafer begins extremely edge fast, then falls to a somewhat center fast condition.

Copper Removal Rate (Integrated) and Platen Temperature as a Function of Polish Time

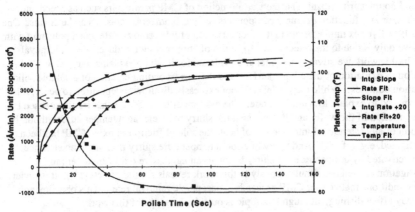

Fig. 1. Integrated copper removal rate as a function of polish time for two positions in the film. Pad surface temperature shows similar behavior. The uniformity is also shown.

The temperature rate relationship was examined under different polishing conditions. A sequence of wafers was polished for a duration adequate to ensure the attainment of thermal stability. The final or equilibrium temperature could then be plotted against the equilibrium removal rate[1]. In the spirit of treating CMP like a chemical reaction, we can plot log of equilibrium removal rate vs. inverse absolute equilibrium (instantaneous) temperature. We have done so for the "standard, two-component" slurry, the abrasive alone and the oxidizer alone, in Fig. 2, below. Note that each of the three sequences results in a straight line when plotted in this manner, and that moreover, the slopes of the three lines are very similar. The implication of the straight line fits (except near room temperature for the "standard" slurry) and the similar slopes is that the polish rate obeys an Arrhenius relationship, and that the activation energy is similar for the slurry and each of its components, differing only in the frequency factor constant. Further, that while the rate for a given slurry proceeds like a chemical reaction, efficacious removal is achieved only with a combination of oxidizer and abrasive.

The different temperatures plotted in Fig. 2 were achieved by varying the level of CMP intensity - i.e. platen speed and downforce. Intensity, therefore is the actual independent variable while both rate and temperature are dependent variables. The observed relationship seen when plotting one against the other may therefore be incidental. To check the effect of an independent introduction of heat into the system, a series of wafers was run using an intentionally elevated platen temperature. In this case, if the reaction rate is truly temperature dependent, the additional data points should fall on the original regression line. If it is not, we should see a line of the same slope shifted leftward from the original. The square data symbols shown on the upper data set are from a wafer sequence polished on an elevated

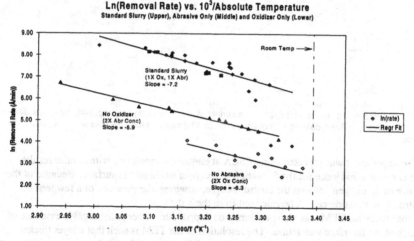

Ln(Removal Rate) vs. 10³/Absolute Temperature
Standard Slurry (Upper), Abrasive Only (Middle) and Oxidizer Only (Lower)

Fig. 2. Log (rate) vs. 1/T for a slurry and its two components (oxidizer and abrasive). A second source of heat is introduced for the square symbols in the "standard" slurry.

---

[1] From the curve of integrated removal rate we can calculate the *instantaneous* removal rate by deconvolution. We are actually interested in the equilibrium value of the instantaneous rate rather than the integrated rate, but after about 60 seconds of polishing, these two are essentially identical. By the nature of the way in which the temperature data was taken, it is always *instantaneous* temperature.

temperature platen. While a 50 °F increase in platen temperature resulted in only a 12 °F increase in pad temperature, it can be seen that the resulting rates fell on the same regression line as the room temperature platen. This implies that the reaction is truly temperature dependent.

The Copper Surface

In addition to rate and temperature observations, the copper surface was thoroughly studied as well to determine if the surface undergoes a transformation during polish. ESCA (XPS), Sputter Auger, and ToF SIMS were employed. Wafers were polished under different conditions and sputter coated with AuPd within an hour after polishing to arrest any further surface oxide growth. Fig. 3 shows two such samples; the as-deposited copper (Fig. 3a) and a sample which was polished then allowed to sit in the oxidizer for two minutes (This would have resulted in a system which was heated at the time the abrasive was cut off and allowed to cool afterward). There is no significant difference in the oxygen concentration near the surface between the two samples. The visible oxygen peak is due solely to exposure of the surface to air, and the tail is due to the depth resolution of the instrument.

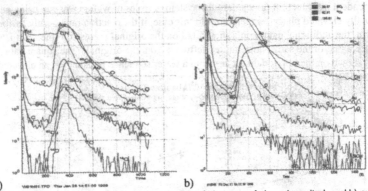

a)   b)

Fig. 3. ToF SIMS counts vs. sputter depth for a variety of elements of a) as-deposited, and b) oxidizer-soaked samples with AuPd coating. An oxygen peak can be seen on each sample at the copper surface.

The experiment was repeated using ESCA at elevated temperature with similar results. There is no clear evidence for the formation of a copper oxide on the surface. Because of the presence of an oxygen peak on the control sample, however, the presence of a few tens of Angstroms of an oxide cannot be ruled out from these data.

Cross sectional TEM was also performed on copper films after polishing. No evidence of an oxide at the interface was found. The resolution of the TEM is such that a layer thicker than about 5Å would be visible, so, as with ESCA, it is not possible to rule out one or two monolayers of converted copper from TEM, but it is possible to rule out the formation of a layer greater than about 10Å.

THE COPPER REMOVAL MECHANISM

The Etch Rate

When evaluating the removal mechanism, it is common to consider the removal rate of the slurry alone, i.e. the etch rate. This can be done separately for the oxidizer and the abrasive and in either case the etch rate is near zero. However we have just shown that the temperature is elevated during polishing, so an accurate assessment of the etch rate can be done only at elevated temperature. The etch rate was measured in the oxidizer for increasing etch times, and in Fig. 4 etch rate as a function of etch time is plotted at 50°C. While the data is subject to a high degree of error, it clearly shows a higher initial rate than would be measured if the measurement were made after one or two minutes. This indicates a strong retarding effect on the etch, such as build-up of a complex on the copper surface, implying that the etch rate is eventually controlled by a dissolution or diffusion rate.

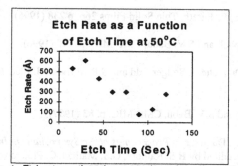

Fig. 4. Etch rate vs. time shows reaction to be self-retarding.

Importantly, the initial etch rate is the rate which is relevant to the CMP process, and it may well be that it is equal to a significant fraction of the observed CMP rate. Further, the actual temperature achieved during polishing is likely to be higher than the equilibrium temperature observed above since it has been observed that the pad cools quickly upon exiting from the carrier[13].

The Model

From all observations in compendium, it is reasonable to model the CMP removal rate for this particular slurry as a *temperature-activated, abrasion-assisted etch process*, in which a by-product of a few monolayers forms to inhibit - but not arrest - the etch process. The abrasive serves the dual purpose of increasing the temperature and clearing the transformed surface to increase the efficacy of the etch. The "oxidizer" is really a controlled etcher/complexer without which the abrasion rate is low. Under normal polishing conditions, the removal rate is limited by the formation of the surface complex. This explains the failure to observe a distinct oxidation layer in TEM as well as the temperature dependence on rate. The model does not require copper redeposition to explain low rate.

ACKNOWLEDGEMENTS

The authors would like to gratefully acknowledge the work and numerous helpful conversations with Chuck Miller, Paul Laberge, J.J. Lee, Jim Conner, Mike Tiner and Janos Farkas.

REFERENCES

1. J. Farkas, D. Watts, J. Saravia, M. Freeman, S. Das, C. Dang, S. Rabke, R. Bajaj, J. Gomez, J. Holley, H. Garcia, K.C. Brooks, M. Wilger, J. Zavala, L. Cook, M. Capetillo, J. Mendonca, J. Klein, Conf. Proc. ULSI XIII, MRS, p. 523 (1998)

2. J.M. Steigerwald, S.P. Murarka, R.J. Gutmann, D.J. Duquette, Mat. Chem. & Phys. 41, p. 217 (1995)

3. R. Carpio, J. Farkas, R. Jairath, Thin Solid Films 266, p 238 (1995)

4. Q. Luo, D.R. Campbell and S.V. Babu, CMP-MIC, p. 145 (1996)

5. C.A. Sainio, D.J. Duquette, J. Steigerwald and S.P. Murarka, J. Elec. Mat., Vol. 25, No. 10, p. 1593 (1996)

6. Q. Luo, M.A. Fury and S.V. Babu, CMP-MIC, p. 83 (1997)

7. C.A. Sainio and D.J. Duquette in *Proceedings of the Symposium On Interconnect and Contact Metallization*, edited by H.S. Rathore, G.S. Mathad, C. Plougonven and C.C. Schuckert (ECS Symp. Proc. Vol. 97-31, 1997) p. 129

8. Q. Luo and S.V. Babu, VMIC, p. 287 (1997)

9. J.M. Steigerwald, S.P. Murarka, J. Ho, R.J. Gutmann and D.J. Duquette, J. Vac. Sci. Tech. B 13(6), p. 2215 (Nov/Dec 1995)

10. J.M. Steigerwald, S.P. Murarka, D.J. Duquette and R.J. Gutmann, Mat. Res. Soc. Symp. Proc. Vol. 337, p. 133 (1994)

11. H-W Chiou, L-J Chen and H-C Chen, CMP-MIC, p 131 (1997)

12. H-W Chiou, Z-H Lin, L-H Kuo, S.Y. Shih, L-J Chen and Chin Hsia, 2nd Annual International Interconnect Technology Conference, May 1999, S.F. CA.

13. R.L. Lane & G. Mlynar, CMP-MIC, p 139 (1997)

ELECTROCHEMICAL BEHAVIOR OF COPPER IN TETRAMETHYL AMMONIUM HYDROXIDE BASED SOLUTIONS

W. H. HUANG*, S. RAGHAVAN*, Y. FANG*, AND L. ZHANG**
* Dept of Material Science and Engineering, University of Arizona, Tucson, AZ
** Now with VLSI Technology, San Jose, CA

ABSTRACT

An investigation was undertaken to characterize the electrochemical behavior of copper in tetramethyl ammonium hydroxide (TMAH) based solutions. The effect of hydrogen peroxide and abrasion with a polyvinyl alcohol (PVA) brush on the corrosion of copper in alkaline solutions were characterized. Galvanic interactions between copper and tantalum in TMAH as well as in ammonium hydroxide (NH$_4$OH) solutions were investigated. The experimental results have shown that the corrosion of copper in TMAH is lower than that in NH$_4$OH, especially at pH values higher than 10. Even in the presence of hydrogen peroxide, TMAH corrodes copper at a lower rate than NH$_4$OH.

INTRODUCTION

Copper is rapidly becoming the interconnect material of choice for devices with feature size of 0.18 μm or below. In the fabrication of such devices, chemical mechanical polishing (CMP) of Cu films is projected to be an integral part in the process scheme. The cleaning of polished copper surfaces surrounding dielectric areas is critical to assure the overall success of device manufacturing. Alkaline solutions based on ammonium hydroxide have been traditionally used in post-CMP cleaning using the double sided scrubbing technique. Since ammonia is a good complexant for copper, there is emerging interest in the use of TMAH based chemicals to reduce the possibility of corrosion of copper. A proprietary acidic solution containing additives has recently been reported to be useful for post-CMP cleaning of copper based structures.[1]

TMAH is a very strong base with a pK$_b$ ~ 0. Unlike ammonia, it does not form complexes with copper. TMAH solutions can also be obtained in very high purity with minimal ionic contamination. The use of TMAH based solution in silicon wafer cleaning has been reported. A cleaning chemical called "Baker Clean" is a TMAH based solution which also contains buffers and proprietary surfactants to remove particles as well as metals after HF cleaning of bare silicon wafers.[2] The use of TMAH-based solutions in post-W-CMP cleaning has been discussed in the literature.[3] Available results show that solutions containing TMAH are very effective in particle removal. However, no information exists as to the effect of TMAH on the corrosion of tungsten.

The objective of this research was to investigate the corrosion behavior of copper in TMAH solutions with and without hydrogen peroxide and compare the results with the corrosion behavior of copper in NH$_4$OH solutions. Since tantalum is most often used as a barrier metal when copper metallization is used, the possibility of galvanic corrosion between copper and tantalum in TMAH solutions was also investigated.

EXPERIMENTAL

The copper films used in the experiments were prepared by electroless plating on a physical vapor deposited (PVD) tantalum film on top of a silicon wafer. The thickness of the copper and tantalum films was 1.5 μm and 300 Å, respectively. Blanket tantalum films were also deposited by PVD on silicon to a thickness of 1000 Å. The samples used for static electrochemical

161

experiments were 1.5 cm × 1.5 cm squares, which were diced from 200 mm wafers. To ensure electrical contact to the backside, all diced samples were coated with nickel print on the backside and along the edge.

The tetramethyl ammonium hydroxide (25% TMAH) used was provided by Mallinckrodt Baker Inc. Both ammonium hydroxide (29% NH_4OH) and hydrogen peroxide (30% H_2O_2) were donated by Olin Microelectronic Materials. In experiments that were conducted in the pH range of 8 to 10, 10^{-3} M potassium nitrate (KNO_3) was added to increase the conductivity of the solution to overcome any IR drop that might occur between the working electrode and the reference electrode.

The electrochemical behavior of copper and tantalum in different solutions was investigated using DC polarization experiments. From the polarization data, the corrosion current density and hence the corrosion rate of copper were calculated using the Stern-Geary Equation[4]

$$i_{corr} = \frac{1}{2.303\,R_p}\left(\frac{\beta_c\beta_a}{\beta_c+\beta_a}\right) \tag{1}$$

where β_a is the anodic Tafel slope, β_c is the cathodic Tafel slope, and R_p is the polarization resistance. The R_p values were calculated from linear polarization data near the corrosion potential. In cases where one of the Tafel slopes was not well-defined (typically β_a), the corrosion current density was calculated by the intersection of the well-defined Tafel slope (typically β_c) at the corrosion potential.

All electrochemical measurements were performed on EG&G Princeton Applied Research 273A Potentiostat/Galvanostat. The Tafel polarization experiments were carried out at a scan rate of 1 mV/sec. The construction of the electrochemical cell used to measure the corrosion rate of copper while being abraded by PVA brush has been described elsewhere[5] (The PVA brush used in this experiment was donated by Cupps Industries, Phoenix, AZ). During the polishing experiments, the down force was controlled at 1.5 psi and the PVA brush was rotated at 100 rpm. The polarization data during abrasion were obtained at a scan rate of 5 mV/sec. The galvanic corrosion experiments were carried out in a flat cell made from PFA and the copper sample was always connected to the working electrode lead of the potentiostat.

RESULTS

Figure 1 displays the corrosion current density (i_{corr}) and corrosion rate of copper in TMAH and NH_4OH solutions as a function of solution pH. In the pH range of 8 to 10, the corrosion of copper is about the same in both TMAH and NH_4OH solutions. At pH values greater than 10, the i_{corr} of copper in NH_4OH solutions increases sharply with increasing pH; the i_{corr} increases from 1 $\mu A/cm^2$ (0.22 Å/min, calculated based on $Cu \rightarrow Cu^{2+}$) at pH~10 to 38 $\mu A/cm^2$ (8.4 Å/min) at pH~11.6. The i_{corr} of copper in TMAH increases only moderately, from 1 $\mu A/cm^2$ (0.22 Å/min) to 9 $\mu A/cm^2$ (2 Å/min), when the pH is increased from 10 to 13.2.

In Figure 2, the effect of H_2O_2 addition to NH_4OH and TMAH solutions on copper corrosion is displayed. The addition of H_2O_2 to the alkaline solutions resulted in a decrease of solution pH. Without any H_2O_2 added to the alkaline solutions at a pH of 8, the i_{corr} of copper is ~1 $\mu A/cm^2$. At the starting TMAH or NH_4OH solution of pH 8, the addition of H_2O_2 results in an increase of i_{corr}; for example at 6% H_2O_2, the corrosion current density increases to ~18 $\mu A/cm^2$ (4 Å/min). If the starting TMAH solution is at pH~12.2, the i_{corr} values increases from 2 to ~18 $\mu A/cm^2$ with increasing H_2O_2 concentration. It is interesting that even though peroxide addition to TMAH increases the corrosion of copper, the increase of corrosion rate is not dependent on the starting pH of the TMAH solution. In contrast, in NH_4OH solutions at a pH ~11.4 without peroxide, the corrosion current density of copper is ~18 $\mu A/cm^2$ (4 Å/min). Even with the

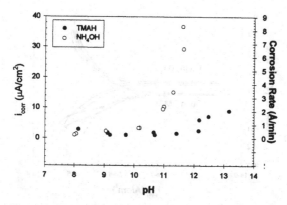

Figure 1: Corrosion of copper as a function of pH in NH₄OH
(O) and TMAH (●) solutions.

addition of 0.3% H_2O_2 to NH₄OH, the i_{corr} of copper increases significantly to ~200 μA/cm² (44 Å/min). Hence in cleaning situations necessitating the use of peroxide containing alkaline solutions, TMAH is preferable to NH₄OH to minimize copper corrosion.

The Tafel polarization plots for copper obtained in TMAH and NH₄OH solutions during abrasion with a PVA pad sample are displayed in Figure 3. A solution pH of 10 was chosen for these investigations. For comparison purposes, Tafel plots obtained before and after abrasion are also plotted in this figure. In both solutions, no significant changes in the i_{corr} or the E_{corr} of copper were seen during abrasion. Based on these results, it may be concluded that light abrasion of copper films with a soft PVA brush is not likely to enhance the corrosion of copper.

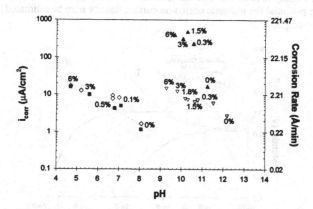

Figure 2: Corrosion of copper as a function of pH in alkaline solutions containing H_2O_2. [The number beside each data point is the H_2O_2 concentration added to the starting alkaline solution. NH₄OH starting pH ~8 (■), TMAH starting pH ~8 (◇), NH₄OH starting pH ~11.2 (▲), and TMAH starting pH ~12.2 (▽).]

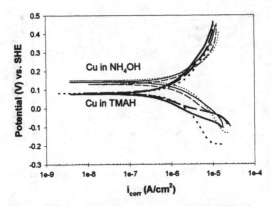

Figure 3: Polarization plots for copper in NH₄OH and TMAH solutions of pH 10 while being abraded by a PVA brush. [(——) during abrasion, (- - - - -) before abrasion, and (— —) after abrasion.]

The possibility of galvanic corrosion between copper and tantalum in TMAH and NH₄OH was analyzed through individual polarization behavior of copper and tantalum measured in these solutions maintained at a pH of 10. Figure 4 and 5 display these polarization curves. In the TMAH solution, copper is more noble than tantalum. The corrosion of tantalum is extremely low in this solution. Additionally, the anodic polarization curve for tantalum intersects the cathodic polarization curve for copper at the corrosion potential of copper; from the intersection point the galvanic potential and galvanic corrosion current density may be read as 100 mV and 10 nA/cm² respectively. Under these conditions, the corrosion of copper is not likely to be influenced by coupling to tantalum. For the coupling of copper and tantalum in NH₄OH (Figure 5), the galvanic potential the galvanic corrosion current density may be estimated to be150 mV and ~30 nA/cm² respectively.

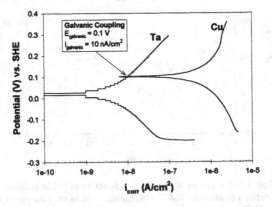

Figure 4: Polarization plots for copper and tantalum in TMAH at pH 10 showing galvanic coupling between the two metals.

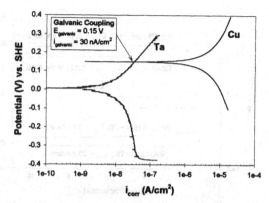

Figure 5: Polarization plots for copper and tantalum in NH₄OH at pH 10 showing galvanic coupling between the two metals.

The estimations of galvanic potentials and current densities were confirmed by experimental measurements shown in Figure 6. In the TMAH solution, the measured galvanic current density is ~ 20 nA/cm² and the potential is ~ 0.125 V. In NH₄OH, the current density is ~ 30 nA/cm² and the potential is ~ 0.15 V. These values agree well with values estimated from the polarizations shown in Figures 4 and 5.

The analysis of galvanic coupling effects based on the plots in Figures 4 and 5 is applicable when copper and tantalum area ratio is equal to one. However, during post-CMP cleaning, the area of tantalum (barrier layer) exposed to cleaning solution would be smaller than the area of exposed copper. For example, if the copper interconnect width is 0.18 μm and the tantalum liner is 0.03 μm thick, the area ratio would be approximately 2.5. Figure 7 illustrates the effect of A_{Cu}/A_{Ta} ratio on the galvanic corrosion between copper and tantalum in TMAH (pH ~10). The increasing area ratio would shift the tantalum polarization to higher corrosion current density. The corrosion of tantalum would increase with increasing A_{Cu}/A_{Ta} ratio, but copper would be increasingly cathodically protected under the same conditions.

CONCLUSIONS

The corrosion of copper in TMAH solutions is less than that in NH₄OH solutions, even at high pH values and high peroxide levels. Closer to neutral pH values, in the absence of peroxide, there are no significant differences in the corrosion of copper in these two types of solutions. Abrasion with a soft PVA brush in the presence of alkaline solutions at a pH of 10 does not appear to enhance the corrosion of copper. Galvanic coupling between copper and tantalum in both TMAH and NH₄OH under conditions where the exposed area of copper is larger than that of tantalum would actually cathodically protect copper.

ACKNOWLEDGEMENTS

The authors would like to thank George Schwartzkopf of Mallinckrodt Baker Incorporated for providing partial funding to carry out this research. Assistance of Joong S. Jeon of AMD (copper and tantalum samples), Nick Andros of Cupps Industries (PVA brush), and Dick Molin of Olin Microelectronics (high purity reagents) is also appreciated.

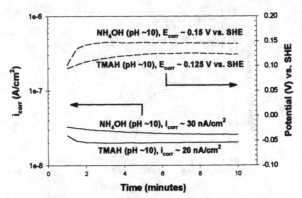

Figure 6: Galvanic corrosion measurements on the copper-tantalum couple in TMAH or NH₄OH at pH of 10. [(——) current density and (— —) potentials.]

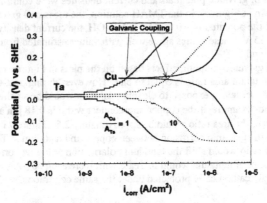

Figure 7: The effect of increasing area ratio between copper and tantalum on galvanic corrosion (TMAH pH 10).

REFERENCES

1. D. Hymes, H. Li, E. Zhao, and J. de Lario, *Semiconductor International*, p. 117-122 (June 1998).

2. W. A. Cady and M. Varadarajan, *J. Electrochem. Soc.*, **143**, p. 2064-2067 (1996).

3. M. Jolley, *Solid State Phenomena*, **65-66**, p. 105-108 (1999).

4. M. Stern and A. Geary, *J. Electrochem. Soc.*, **104**, 56 (1957).

5. E. A. Kneer, C. Raghunath, V. Mathew, S. Raghavan, and J. S. Jeon, *J. Electrochem. Soc.*, **144**, p. 3041-3049 (1997).

PROCESS CONTROL CHALLENGES AND SOLUTIONS: TEOS, W, AND CU CMP

J. Mendonca*, C. Dang, S. Selinidis, M. Angyal, B. Boeck,
R. Islam, C. Pettinato, P. Grudowski, J. Cope, B. Smith, and V. Kolagunta
Advanced Products Research and Development Laboratories, Motorola, Austin TX
*Author to whom correspondence should be addressed

ABSTRACT

Various options that afford control of the TEOS, W, and Cu/barrier polishes were explored in the building of multilevel dual inlaid structures. Improved tool performance that enables more sophisticated down pressure control with higher resolution backpressure adjustments was employed for the oxide module to control the interlevel capacitances. Planarity at both the global and local levels at the oxide polish affords a good starting point for successive builds without metal pooling. In W CMP, small and controllable oxide erosion and plug recess was obtained with harder polishing pads. In Cu/barrier CMP, the tight overpolish/underpolish margin was maintained by head control and appropriate endpoint algorithms. A six-level build with tight and low sheet resistances and leakages was demonstrated.

INTRODUCTION

In inlaid logic devices there is a move away from oxide to metal CMP for more levels of build. However, the process control required for the oxide/W polish levels is tighter because of the requirement of a high degree of planarity from the upper inlaid metal layers. For Cu, a tight polish window with controlled over/under polish has to be maintained at all metal levels for a successful multilevel structure build.

EXPERIMENTAL

All the tests were carried out on 8" wafers on commercially available one head dual platen and three head four platen CMP tools. The film rates and uniformities were measured on a Tencor UV1250 and a four point probe RS55. For this set of experiments, a KOH-based slurry was used for TEOS polish, ferric nitrate based for W, and a proprietary slurry for Cu/barrier CMP. The monitoring of the oxide erosion was done by oxide thickness and profile measurements. In addition, electrical measurements were conducted on selected patterns to monitor leakage and contact resistance. With W CMP, oxide erosion tests were conducted on two device types (Logic and SRAM), four different mask sets with successively finer technology design rules, viz. 0.25μm, 0.18μm, and 0.13μm. Across wafer pattern density effects could be evaluated at different line widths. On the 0.18μm and 0.13μm sets, three different pattern densities at the local interconnect level were evaluated. The first, called dens1 (contact chains), had the lowest pattern density. The second, called dens2 (local interconnect pattern), had an intermediate density level; while the third, called dens3 (process control structures comprised of large serpentine combs at the local interconnect level), had the highest pattern density.

RESULTS AND DISCUSSION

In oxide CMP, the greatest challenge is to have a uniform post-polish thickness within and across wafers. Figure 1 shows a cross-section schematic after polish. It is simplified to display the effects of CMP on electrical parameters. With oxide polish, maintaining a uniform post-CMP thickness within and across wafers ensures that one would sustain a tight distribution of interlevel capacitances. Various strategies can be employed for reining in the non-uniformity. The incoming film should be controlled for thickness and dopant concentration uniformity because the pre-polish non-uniformities affect the post-CMP result strongly. In the polish process itself, head or carrier backpressures can be tuned to affect the center to edge rate. Our backpressure adjustments resulted in an acceptable planarity. This tight process control enables one to obtain repeatable and within specification interlevel capacitances. Oxide topography has to be planar at both the local and global level. Global variations can be corrected by intentional non-uniformity compensation at the metal polish steps. Setting the

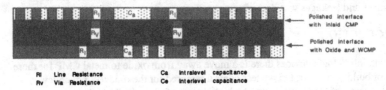

| RI | Line Resistance | Ca | Intralevel capacitance |
| Rv | Via Resistance | Ce | Interlevel capacitance |

Figure 1 Polish effect on electrical parameters

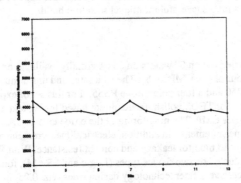

Figure 2a Post-CMP oxide thickness across wafer on different sites

topography correctly at the interlevel dielectric oxide level is critical because no self-planarizing occurs at the backend. However, local planarity can create issues at the higher levels as evidenced by metal pooling. The device structure used for these experiments does not allow for capacitance measurements. The planarity is evaluated, instead, with oxide metrology. Figure 2 shows the remaining oxide thickness on dies across a typical wafer measured on 13 sites. The 3 sigma range was 507A. At each site, the variation across the die ranged from 100 to 200A.

In W CMP, good planarity implies small amounts and controllable oxide erosion and plug recess. From Figure 1, we can see that if one controls the planarity and the oxide erosion, then one can obtain within-specification via resistance and interlevel capacitance. Several options were explored to limit the oxide erosion on all of the widely varying density structures. Consumable and tool parameter variations showed the most promise. The result of such optimization is shown in Figure 3. The pad types that were tried: 1. softer polyurethane based, harder polyurethane based, and a harder polyurethane based with special grooves for slurry transport. The results reiterate that the oxide erosion increase with higher pattern densities, more variation of pattern densities, and lower pad hardness. Cross-sections after W CMP show good planarity as depicted in Figure 4. The tight sheet resistance (Figure 5) on a 0.175 micron by 3.5 micron 4 terminal resistor after polish indicates that there is acceptable oxide dishing and erosion. The hard pad process allows for a sufficiently planar foundation which in turn enables multiple level backend build. As indicated in Figure 4, increasing the pad hardness decreases the maximum across wafer planarity to the 600A range thus allowing higher level build.

The challenge with Cu/barrier CMP is that the underpolish/overpolish is tight. With an underpolish one would encounter leakage with residual barrier/Cu, while with overpolish high sheet resistances on the large bus lines result due to excessive dishing. Also, the overpolish margin is limited by the possibility of excessive oxide erosion. Thus, the line resistances and both inter- and intra-level capacitances would be impacted as shown in Figure 1. In addition, the inability to maintain adequate polish control at an underlying level would set the stage for excessive residual and consequent scrap at the higher levels (please see Figure 6). Improved tool performance that enables higher resolution head control can be utilized to control the within and across wafer non-uniformities. In Figure 7, we can see the tight polish margin in terms of overpolish required in order to meet the sheet resistance specification on a 0.28micron spaced 22.5 micron wide dummy lines. One way to control the overpolish/underpolish extent is to focus on developing a robust endpoint algorithm. The results from using one such algorithm is displayed in Figure 8. The repeatability of the endpoint triggering was held to less than 4% 1-sigma. Note that this includes and takes account of the incoming film thickness variation. Figure 9 shows the results that were obtained on sheet resistance and leakage. Such tight control of across and within wafer non-uniformity and optimized endpoint triggering allows for within specification build of multilevel dual inlaid devices. A six-level cross-section (Figure 10) demonstrates the successful build utilizing the controlled TEOS, W, and Cu/barrier CMP processes.

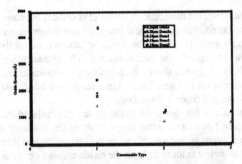

Figure 3 Oxide erosion versus consumable type

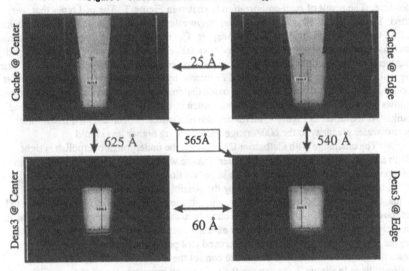

Figure 4 SEM cross-section of cache and Dens3 structure on a 0.13μm vehicle on a hard pad

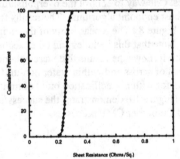

Figure 5 Local interconnect sheet resistance

Figure 6 Inadequate process control at underlying level

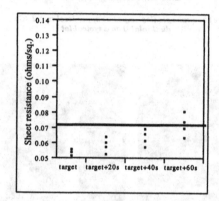

Figure 7 Overpolish tolerance

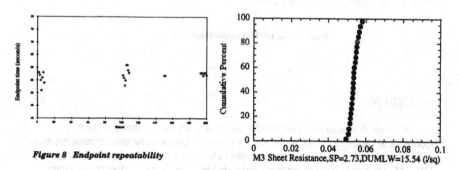

Figure 8 Endpoint repeatability

Figure 9a Sheet resistances (ohms/sq.) at upper level metal Cu dual inlaid on a typical lot

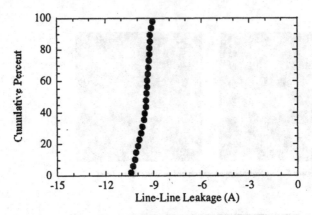

Figure 9b   Leakages (log amps.) at upper level metal   (Cu dual inlaid) on a typical lot

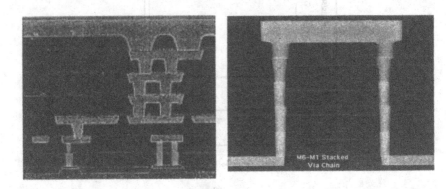

Figure 10 Six levels of copper integrated

CONCLUSIONS

We have demonstrated a successful six-level build after optimizing the unit TEOS, W, and Cu/barrier CMP processes. Within specification electrical parameters and repeatability of the process results were shown to be adequate for the dual inlaid logic and SRAM structures. Tool parameters, consumable changes, and endpoint algorithms were identified as key in the TEOS, W, and Cu/barrier CMP processes studied.

ELECTRICAL CHARACTERIZATION OF SLURRY PARTICLES AND THEIR INTERACTIONS WITH WAFER SURFACES

Jin-Goo Park, Sang-Ho Lee, Hyoung-Gyun Kim, Department of Metallurgy and Materials Engineering, Hanyang University, Ansan, 425-791, Korea; Hea-Do Jeong, Department of Mechanical Engineering, Pusan National University, Pusan, 609-735, Korea; Doo-Kyung Moon, Hanwha Group R&D Center, Taejon, 305-345, Korea

ABSTRACT

The purpose of this study was to explore the interaction between slurry particles and wafer surfaces by the measurements of their zeta potentials. The zeta potentials of slurry particles such as fumed and colloidal silica, alumina, ceria and MnO_2 and substrates such as silicon, TEOS, W, and Al have been measured by electrophoretic and electroosmosis method to evaluate the electrical properties of surfaces, respectively. The zeta potential of oxide and metal surfaces showed similar values to those of particles as a function of pH. The interaction energy between alumina and silica particles and TEOS, W and Al substrate were calculated based on DLVO theory. No deposition of silica particles on TEOS and the heavy deposition of alumina particles on metal substrates were observed in the particle deposition test. Experimental results were well agreed with the theoretical calculation.

INTRODUCTION

The interaction force between slurry abrasives and surfaces during CMP is mainly determined by van der Waals attractive and electrostatic repulsive force [1]. Even though van der Waals force is inherent in materials, the electrostatic forces are strongly dependent on the potentials on surfaces which are strongly dependent on the pH of solutions. The surface potentials of particles and wafer surfaces in an aqueous solution can be expressed in terms of zeta potentials. If the sign of potentials of two interacting surfaces are the same, there is a net repulsion of two surfaces and vice versa depending on the magnitude of van der Waals force.

DLVO theory [2,3] can be used to calculate the interaction forces between the slurry particle and the wafer surface to be polished. The interaction forces between particles and between particles and surfaces could provide important information on the stability of slurry and the degree of particle contamination on surfaces after CMP.

Post CMP cleaning is a very important process to accomplish the main purpose of planarization. Particles in slurry are usually left on polished surfaces and can be removed by either contact [4] or non-contact [5] mode cleanings. Metal CMP has been known to leave severe particles and metallic contaminants on polished surfaces when compared to oxide CMP.

In this paper the zeta potentials of slurry particles and wafer surfaces to be polished were measured as a function of solution pH and the interactions forces between them were calculated to correlate the level of contamination after CMP. The deposition of particles were also performed to verify the theoretical calculation.

EXPERIMENT

For the experiment, bare silicon, thermal oxide (600 nm) and TEOS (1.5 μm) coated silicon, Al (500 nm) and W (700 nm) deposited wafers were used as substrates. Al was sputter coated on SiO_2 and W was prepared by CVD on Ti/TiN (20/60 nm) layers. Substrates were cut into 15 mm × 30 mm for the zeta potential measurements and pre-cleaned before the

173

measurements. Si and oxide substrates were cleaned in the mixture of H_2SO_4 and H_2O_2 (4:1) followed by HF treatment and finally rinsed with high purity deionized (DI) water (18.2 MΩ·cm, manufactured by Millipore Milli-Q Plus). Al substrate was slightly etched in the solution mixture of 73% H_3PO_4, 4% HNO_3, 3.5% CH_3COOH, and 19.5% DI water for 1 min 30 sec and rinsed with DI water. W was etched in the solution mixture of 3.4 % KH_2PO_4, 1.34% KOH, 3.3% $K_3Fe(CN)_6$ and DI water for 2 min. All samples were N_2 dried after cleanings.

γ-alumina of 0.05 μm and colloidal silica of 0.06 μm in diameter were purchased from Buehler Co. Fumed silica of 0.014 μm was obtained from Sigma Co.. Ceria and MnO_2 particles were provided by Mitsui Mining and Smelting Co. W particles of -325 mesh size were purchased from Cerac. Also commercial oxide and metal slurries were analyzed for the evaluation and comparison purpose.

The zeta potentials of slurry particles and substrates have been measured by electrophoretic and electroosmosis method to evaluate the electrical properties of surfaces, respectively, by using a LEZA-600 Laser Electrophoresis Zeta Potential Analyzer of Otsuka Electronics Co. The zeta potential of particles and plate was measured in $10^{-3}M$ KCl and $10^{-2}M$ NaCl solution, respectively. The zeta potential of plate was measured by using a proprietary monitor solution provided by Otsuka Electronics Co. pH and redox potentials of slurry solutions were measured by a Orion's 520A pH meter with Ag/AgCl electrode and a Pt redox electrode. The surface tension of solutions was measured by a Cahn's dynamic contact angle analyzer (DCA 315). Particle deposition experiments were performed to verify the theoretical interaction energy calculation between particles and wafers. Particles were deposited on substrates by dipping them in slurry contained beaker for 3 min, then dried in air. The magnitude of particle deposition was observed by a dark filed of an optical microscope. pH was controlled with HCl and NaOH.

RESULTS AND DISCUSSION

The selection of slurry abrasives is one of the most important task in CMP process development. It will determine the removal rate and the level of defects such as particles and scratches. In this study various slurry particles and surfaces to be polished were chosen to measure their electrical properties in aqueous solutions. The harder particles, the greater the removal rates. Table 1 shows the hardness of materials of interest to CMP process. Among particles in Table 1, γ-alumina, CeO_2, MnO_2, fumed and colloidal silica particles were used to measure their zeta potentials as a function of solution pH.

Table 1. Hardness of materials related to CMP [6]

Materials	Hardness (Mho's scale)	Materials	Hardness (Mho's scale)
Al_2O_3 (α)	9	MnO	4.7 - 5.6
Al_2O_3 (γ)	8	MnO_2	2 – 6
CeO_2	6	SiC	9.5
Cu_2O	3.5 – 4	SiO_2	6 – 7
MgO	5 - 6.5	WO_2	5.5 – 6

The zeta potential of slurry particles was measured as shown in Figure 1. Fumed silica showed a higher isoelectric point (IEP) at which the net charge and electrophoretic mobility is zero, than that of colloidal silica as shown in Figure 1 (a). Also zeta potentials of colloidal silica were around 20mV lower than that of fumed silica. Figure 1 (b) shows the zeta potentials of alumina, ceria and MnO_2 particles. Due to their lower hardness than alumina, ceria has been widely used in polishing industry [7]. Recently MnO_2 has been introduced in metal CMP because it would dissolve in a very dilute solution of hydrogen peroxide, water and an acid [8]. It was claimed that the application of MnO_2 particles in CMP would reduce the particle contamination after CMP. The highest IEP of 9.8 was measured in alumina particles. MnO_2 particles showed the IEP at near pH 6. Ceria particles showed the negative zeta potentials in the pH ranges investigated. It should be noted that the preparation of slurry in acidic pH ranges would cause the unstable suspension of silica and ceria particles based on the zeta potential measurements of particles.

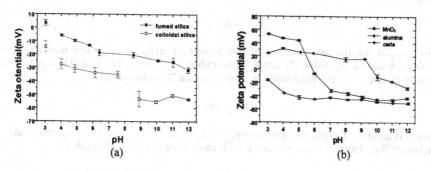

Figure 1. Zeta potentials of (a) fumed and colloidal silica and (b) alumina, ceria and MnO_2 particles as a function of solution pH.

Figure 2 shows the zeta potentials of wafer surfaces of interest to oxide and metal CMP. Aluminum surface showed its IEP at around pH 8.5 which was lower than that of alumina as shown in Figure 2(a). However, zeta potential of Al was very similar to that of alumina particles as a function of pH. TEOS wafer also showed a very similar zeta potentials to those of colloidal silica even though there was a slight shift to lower negative potentials. Also the zeta potential of bare silicon surface was measured at different pHs and the IEP was slightly larger than 3. Silicon surface has a zeta potential lower than –80 mV above pH 9.

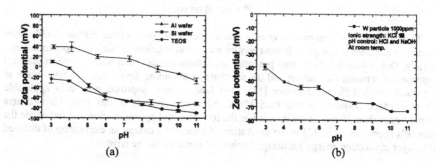

Figure 2. Zeta potentials of (a) silicon, TEOS and aluminum wafers and (b) W particles as a function of pH

Even though W surfaces were tried to measure their zeta potentials as a function of pH, it was not able to measure them due to the low cell constant induced by the high surface conductivity of metal. Instead, zeta potentials of W particles were measured as shown in Figure 2 (b) and used for the interaction force calculations.

DLVO [2,3] theory estimates the repulsive and attractive force due to the overlap of electric double layers and London-van der Waals force in terms of inter particle distance, respectively. The summation of them gives the total interaction force and can be used for the interpretation of colloid stability in terms of the nature of interaction force–distance curve. If a small interparticle separation (H) is assumed, van der Waals forces for a sphere and substrate can be expressed to

$$V_A = - A_{132} R/6H \tag{1}$$

where, A_{132} is the Hamaker constant of two particles 1 and 2 in dispersion medium 3 and calculated based on A_{11} value [1] and R is the radius of a sphere. The Hamaker constants of unknown metals were calculated based on their definition [9] as shown below;

$$A_{11} = \pi^2 n^2 C_1 \tag{2}$$

where, n is number of atoms per unit volume, C_1 is $(3/4) h\nu_0 \alpha^2$, $h\nu_0$ is ionization potential and α is polarizability. Hamaker constants used for this study were summarized in Table 2.

Table 2. Hamaker constants used for van der Waals force calculation.

Materials	$A_{11} \times 10^{-20}$ J
Si	25.6 [10]
SiO$_2$	50 [10]
Al$_2$O$_3$	15.5 [10]
Water	43.8 [10]
W	465*
Al	179*

* = based on eq. (2)

The calculation of interaction energy due to the overlapping of the diffuse double layer between two surfaces is complex. It must rely on numerical solutions or various approximations. Generally, Debye-Huckel (D-H) low potential approximation has been widely used to interpret the interaction between the particle and the substrate by assuming the particle and flat surface as two particles with different diameter [11]. D-H low potential approximation only applicable when the surface potential is less than 25mV. Most of materials in this study showed zeta potential larger than 25mV. If we do not use the low potential approximation and rather use the Overbeek's approximation [12] in the derivation of interaction energy an expression of diffused double layer interaction energy for unequal spherical particles can be written as;

$$V_R = \frac{64\pi\varepsilon a_1 a_2 k^2 T^2 \gamma_1 \gamma_2}{(a_1 + a_2)e^2 z^2} \exp[-\kappa H] \tag{3}$$

where z is the counter-ion charge number, a_1 and a_2 are particle radius, κ is $(8\pi n \nu^2 e^2/\varepsilon k T)^{1/2}$ and

$$\gamma = \frac{\exp[ze\,\psi_d\,/2kT]-1}{\exp[ze\,\psi_d\,/2kT]+1} \tag{4}$$

The total interaction energy was calculated between silica and alumina particles and TEOS, W and Al wafer surfaces at pHs which were interested in oxide and metal CMP. Figure 3 shows the total interaction energy between colloidal and alumina particles and TEOS wafers at pH 11. The application of the low potential approximation to the calculation of V_R resulted in the attractive total interaction energies as a function of interparticle space regardless of particles and substrates due to the dominant van der Waals energy term. However V_R based on Overbeek's approximation resulted the repulsive total interaction energy which was well agreed with the particle deposition experiments. In the particle deposition experiments, no silica and alumina particles were deposited on TEOS wafers at pH 11.

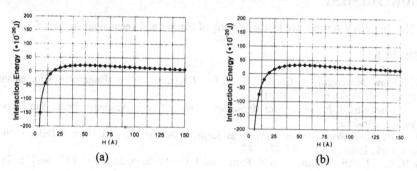

Figure 3. DLVO total interaction energy between TEOS surface and (a) colloidal silica and (b) alumina particles at pH 11.

Figure 4 shows the total interaction energy between metal surfaces and alumina particles at pH 4 and 11. Particle dip tests showed the heavy deposition of particles on metal surface at both pHs. The magnitude of deposited particles was greater at pH 4. As shown in Figure 4, the strong adhesion of alumina particles on metal surfaces were calculated.

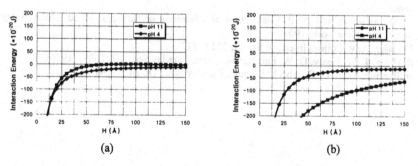

Figure 4. DLVO total interaction energy between alumina particles and (a) Al and (b) W substrates at pH 4 and 11.

CONCUSIONS

Zeta potentials of slurry particles and wafer surfaces were measured to calculate the DLVO total interaction energy between them at various pHs. Instead of the Debye-Huckel low potential approximation, Overbeek's approximate was applied to the calculation. The repulsive energy was calculated between silica and TEOS wafers. Particle dip test also showed no deposition of particles on TEOS wafer. Due to the low cell constant of conductive W plate, it was not possible to measure the zeta potentials of W. The Hamaker constants of Al and W were calculated and applied to the calculation of total interaction energy. The theoretical calculation was agreed well with the experimental results. The strong attractive interaction between metal surfaces and alumina particles were observed in both the calculation and experiments.

ACKNOWLEGEMENT

This research was supported by the Ministry of Science and Technology through STEPI.

REFERENCES

1. D. J. Shaw, *Introduction to Colloid and Surface Chemistry*, 4th ed. (Butterworth-Heinemann Ltd.,Oxford, 1992), p. 174-209, 214.
2. J. N. Israelachvili, *Intermolecular & Surface Forces*, 2nd ed. (Academic Press, San Diego, 1992), p. 246-256.
3. P. C. Hiemenz, *Principles of Colloid and Surface Chemistry*, 2nd ed. (Marcel Dekker INC, New York, 1986), p. 686-697, 710-727.
4. S. R. Roy, L. Ali, G. Shinn, and N. Furusawa, J. Electrochem. Soc. Vol. 142, No.1, p. 216 (1995).
5. A. A. Busnaina, I. I. Kashkoush, and G. W. Gale, J. Electrochem. Soc. Vol. 142, No. 8, p. 2812 (1995).
6. G. V. Samsonov, *The Oxide Handbook*, 2nd ed. (IFI/Plenum Data Company, New York, 1982), p.192.
7. Y. Homma, T. Furusawa, K Kusukawa, M. Nagasawa, Y. Nakamura, M. Saitou, H. Morishima, H. Sato, *Proceedings of 1995 VMIC Conference*, p. 457.
8. K. Hanawa, *Proceedings of Int'l CMP Technical Symposium 98*, SEMICON Korea 98, Seoul, p.163.
9. A. W. Adamson and A. P. Gast, *Physical Chemistry of Surfaces*, 6th ed. (John Wiley & Sonic, Inc., New York, 1997), p.232.
10. J. Visser, Adv. Colloid Interface Sci., 3, 331 (1972).
11. R. Hoog, T. W. Healy, D. W. Fuerstenau, Trans. Faraday Soc., 62, 1638, (1996).
12. H. Reerink, J. Th. G. Overbeek, Discuss. Faraday Soc., 18, 74 (1954).

Part V

CMP Modeling and
Fluid Flow

Hydrodynamics of a Chemical-Mechanical Planarization Process

IN-SUNG SOHN[1], BRIJ MOUDGIL[2], RAJIV SINGH[2], C.-W. PARK[3]
[1,3]Department of Chemical Engineering, University of Florida, Gainesville, FL 32611
[2]Department of Material Science and Engineering, University of Florida, Gainesville, FL 32611

ABSTRACT

The flow of a slurry in a CMP process has been investigated. This wafer-scale model provides the three dimensional flow field of the slurry, the spatial distribution of the local shear rate imposed on the wafer surface, and the streamline patterns which reflect the transport characteristics of the slurry.

INTRODUCTION

As the devices become more miniaturized in microelectronics industry, the required level of planarization calls for a more stringent thickness control, and the CMP process is expected to meet the planarity requirements.

Modeling of the CMP process is often classified into two categories; wafer-scale model and feature-scale model. The characteristic length scale of the wafer-scale model is the gap between the pad and wafer which is in the order of 50 μm, and it attempts to describe the overall removal rate of the CMP process. The feature-scale model is for the length scale of typical device features on the wafer which is in the order of a few micrometers, and focuses on the local removal rate rather than the overall removal rate.

The wafer-scale model which is most frequently referred to is Preston's equation[1]. According to this equation, the average removal rate is proportional to the applied normal force (or pressure) and the speed of the wafer relative to that of the pad. In another wafer-scale model, Patrick et al.[2] found that every point on the wafer experiences the same linear speed relative to the pad if the wafer rotates with the same angular velocity as the pad. Their experiment confirmed that the variation of the local removal rate was affected by the relative angular velocity of the wafer and applied normal force (or pressure). Ouma et. al.[3] showed that the polishing rate near the edge of the wafer is higher than the center and this can be attributed to the reduced slurry flow at the interior of the wafer and the increased stress at the wafer edge. Runnel et al.[4] proposed a model which accounts for the fluid flow between the wafer and the pad. Wang et al.[5] presented a model correlating the non-uniformity with the Von Mises stress distribution, which is a combination of principal stresses. Several other studies attempted to account for various effects of slurry flow and the slurry-pad interaction on the removal rate and the non-uniformity across the wafer[4,7].

The present study is a wafer-scale modeling which provides the three dimensional flow field of the slurry in the gap between the wafer and the pad which are assumed to be parallel to one another.

GOVERNING EQUATIONS

A schematic of a CMP process is shown in Figure 1. A circular wafer of a radius R_w rotates with an angular velocity Ω_w about its center, whereas the pad with a much larger radius rotates with an angular velocity Ω_p and its center of rotation is a distance L away from the center of the

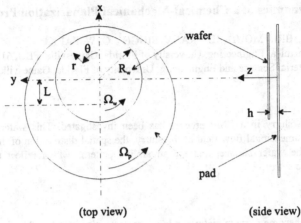

(top view)   (side view)

Figure 1. Schematic of a CMP process

wafer. Assuming that the pad and the wafer are rigid and parallel with a uniform gap of h and that the slurry in the gap is a Newtonian fluid, the governing equation for the slurry flow in the gap is given as

$$\nabla \cdot \mathbf{u} = 0 \tag{1}$$

$$\rho(\mathbf{u} \cdot \nabla \mathbf{u}) = -\nabla p + \mu \nabla^2 \mathbf{u} \tag{2}$$

Since the fluid motion is induced by the relative motion of the two parallel plates (i.e., wafer and pad) as shown in Figure 1, the flow is purely a shear flow and consequently, the velocity component in the gap direction is zero and the isotropic pressure is constant. Thus, the governing equation in a cylindrical coordinate system shown in Figure 1 is reduced to

$$\frac{1}{r^*}\frac{\partial}{\partial r^*}(r^* u_r^*) + \frac{1}{r^*}\frac{\partial u_\theta^*}{\partial \theta} = 0 \tag{3}$$

$$Re\,\varepsilon^2\left(u_r^*\frac{\partial u_r^*}{\partial r^*} + \frac{u_\theta^*}{r^*}\frac{\partial u_r^*}{\partial \theta} - \frac{u_\theta^{*2}}{r^*}\right)$$
$$= \left[\varepsilon^2\frac{\partial}{\partial r^*}\left(\frac{1}{r^*}\frac{\partial}{\partial r^*}(r^* u_r^*)\right) + \frac{\varepsilon^2}{r^{*2}}\frac{\partial^2 u_r^*}{\partial \theta^2} + \frac{\partial^2 u_r^*}{\partial z^{*2}} - \frac{2\varepsilon^2}{r^{*2}}\frac{\partial u_\theta^*}{\partial r^*}\right] \tag{4}$$

$$Re\,\varepsilon^2\left(u_r^*\frac{\partial u_\theta^*}{\partial r^*} + \frac{u_\theta^*}{r^*}\frac{\partial u_\theta^*}{\partial \theta} + \frac{u_r^* u_\theta^*}{r^*}\right)$$
$$= \left[\varepsilon^2\frac{\partial}{\partial r^*}\left(\frac{1}{r^*}\frac{\partial}{\partial r^*}(r^* u_\theta^*)\right) + \frac{\varepsilon^2}{r^{*2}}\frac{\partial^2 u_\theta^*}{\partial \theta^2} + \frac{\partial^2 u_\theta^*}{\partial z^{*2}} + \frac{2\varepsilon^2}{r^{*2}}\frac{\partial u_r^*}{\partial \theta}\right] \tag{5}$$

Appropriate boundary conditions are the no-slip condition at the pad ($z^*=0$) and the wafer ($z^*=1$) surfaces:

At $z^* = 0$; $u_r^* = \omega\,\ell\sin\theta$, $u_\theta^* = \omega(r^* + \ell\cos\theta)$ \hfill (6a,b)

At $z^* = 1$; $u_r^* = 0$, $u_\theta^* = r^*$ \hfill (7a,b)

The above governing equations and boundary conditions are in dimensionless form. The superscript * indicates the dimensionless variables. Appropriate scalings for the non-dimensionalization are the wafer radius R_w and the gap thickness h for the r- and z-coordinate, respectively. Obviously, the θ-directional component of the velocity should be scaled by $R_w\Omega_w$. Since the two terms in the continuity equation (Equation (3)) should be balanced, the scale for u_r should be also $R_w\Omega_w$. The four dimensionless parameters are appearing in these equations are

$$\text{Re} = \frac{\rho R_w^2 \Omega_w}{\mu} \quad , \quad \varepsilon = \frac{h}{R_w} \quad , \quad \ell = \frac{L}{R_w} \quad , \quad \omega = \frac{\Omega_p}{\Omega_w}$$

RESULTS AND DISCUSSION

In a typical CMP process, the orders of magnitude of Re and ε are 10^4 and 10^{-4}, respectively. Consequently, $\text{Re·}\varepsilon^2$ and ε^2 are as small as $O(10^{-4})$ and $O(10^{-8})$. Thus the terms associated with these small quantities are negligible, and the leading order solution to Equations (4) and (5) which satisfies the boundary conditions (6) and (7) can be obtained as

$$u_r^* = \omega \, \ell \sin\theta \left(1 - z^*\right) \quad \text{and} \quad u_\theta^* = \omega \, (r^* + \ell\cos\theta)(1 - z^*) + r^* z^* \qquad \text{(8a, b)}$$

The order of magnitude of the error in this solution under the given assumptions is $O(\text{Re·}\varepsilon^2, \varepsilon^2)$. With the velocity given as above, the shear rate is calculated as $\dot\gamma = \sqrt{2\mathbf{D}{:}\mathbf{D}}$ by the definition where the rate of deformation tensor $\mathbf{D}$ is $\frac{1}{2}\left\{\nabla u + (\nabla u)^{\mathbf{T}}\right\}$. Appropriate scale for the shear rate is $R_w\Omega_w/h$. Thus, the dimensionless shear rate $\dot\gamma^{*}$ is given as

$$\dot\gamma^* = \left[\left\{r^* - \omega(r^* + \ell\cos\theta)\right\}^2 + (\omega\ell\sin\theta)^2\right]^{\frac{1}{2}} \qquad (9)$$

Here the scale for the $\dot\gamma^{*}$ is $R_w\Omega_w/h$.

Since every point on the wafer surface with the same radial position r experiences the same average shear rate for each revolution, more appropriate representation of the local shear rate is the average value over θ:

$$\dot\gamma^*_{avg} = \frac{1}{2\pi}\int_0^{2\pi}\left[\left\{r^* - \omega(r^* + \ell\cos\theta)\right\}^2 + (\omega\ell\sin\theta)^2\right]^{\frac{1}{2}}d\theta \qquad (10)$$

In Figure 2, the dependence of the average shear rate on the radial position is shown for various values of ω ($=\Omega_p/\Omega_w$) when $\ell=0.5$. This result suggest that the average surface removal rate is increasingly greater toward the outer edge of the wafer unless the angular velocity of the pad is same as that of the wafer. When $\omega=1$ (i.e., $\Omega_p = \Omega_w$), the dimensionless shear rate is uniform throughout the wafer surface with a magnitude of ℓ (which is equivalent to the dimensional value of $\Omega_w L/h$). Thus, in the CMP process where the pad and the wafer are rotating eccentrically, it is preferable to maintain their rotational speed to be the same in order to achieve a uniform distribution of the shear rate throughout the wafer surface. This result was previously pointed out by Patrick et al.[2]

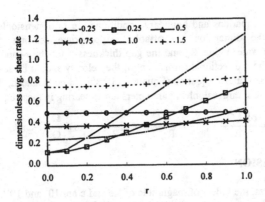

Figure 2. Radial dependence of the dimensionless average shear rate $\dot{\gamma}^*_{avg}$ for $\omega = -0.25 \sim$ 1.5 when $\ell=0.5$ (The scale for the shear rate is $R_w\Omega_w/h$, and the average is over the angle θ.)

One of the most important factors which influence the CMP process is the local transport of the slurry, and the streamlines of the flow inside the gap between the pad and the wafer may provide an insight into it. Since the velocity field is two-dimensional as given in Equation (8), the stream function ψ can be determined as

$$\psi = r^*\omega(z^*-1)\left(\frac{1}{2}r^* + \ell\cos\theta\right) - \frac{1}{2}r^{*2}z^*$$ (11)

Here the scale for the dimensionless stream function ψ is $R_w^2\Omega_w$, and ψ has been set to zero at $r^*=0$ without loss of generality. Streamlines are represented by the contours of constant ψ, and the streamlines at a z^*=constant plane are shown to be concentric circles with their center at (x, y) = $(-\omega\ell(z^*-1)/\{\omega(z^*-1)-z^*\}, 0)$. Several streamlines at a plane with different value of z^* are shown in Figure 3. Since the pad and the wafer are rotating eccentrically, some streamlines are cutting across the outer rim of the wafer whereas some are closed circles within the wafer rim boundary (Figure 3b or 3c). The slurry inside the outer-most closed streamline will stay within that region whereas the fluid outside of it will be replaced by freshly fed slurry as the pad rotates. For the purpose of the surface planarization, the presence of the stagnant region is not desirable and it should be minimized.

The radius of the outer-most close streamline R_c is $1-[\omega\ell(z^*-1)/\{\omega(z^*-1)-z^*\}]$ and the volume of the stagnant region relative to the total volume of the gap between the pad and the wafer (V_s/V) is

$$\frac{V_s}{V} = \int_0^1 R_c^2 dz^* = \int_0^1 \left\{1 - \frac{\omega\ell(z^*-1)}{\omega(z^*-1)-z^*}\right\}^2 dz^*$$ (12)

As pointed out previously, ω should be 1 for the uniform shear rate throughout the wafer surface. In this case (i.e., when $\omega=1$),

$$\frac{V_s}{V} = \begin{cases} 1-\ell+\dfrac{1}{3}\ell^2 & \text{when } \ell\leq 1 \\[2mm] \dfrac{1}{3\ell} & \text{when } \ell>1 \end{cases}$$ (13)

184

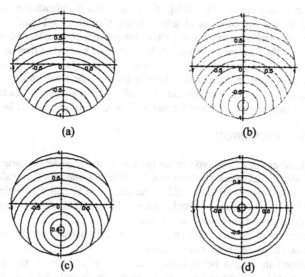

Figure 3. Streamlines at constant-z planes for $\ell=1.0$ and $\omega\ (=\Omega_p/\Omega_w) = 1.0$
(a) $z^*=0$ (pad surface), (b) $z^*=0.25$, (c) $z^*=0.5$, (d) $z^*=1$ (wafer surface)

When $\ell=0$, the center of the pad matches with that of the wafer and the entire volume of the gap is stagnant (i.e., $V_s/V = 1$). As ℓ increases, V_s/V decreases monotonically. When $\ell=1$, V_s/V is 1/3 meaning that one third of the gap space below the wafer is a stagnant region. When $\ell > 1$, the center of the pad is outside the wafer rim and consequently, the sweeping motion of the pad is more efficient and V_s/V is smaller than 1/3.

The size of the stagnant region can be reduced by increasing ℓ as indicated in Equation (13). However, increasing ℓ is costly since it requires a bigger pad. Although the result of the integral in Equation (12) is not shown due to its algebraic complexity, larger $\omega\ (=\Omega_p/\Omega_w)$ makes V_s/V smaller. As it was pointed out previously (Figure 2), however, the spatial non-uniformity of the shear rate becomes significant if ω deviates from 1. Thus, increasing the value of either ℓ or ω is not a practical approach to reduce the size of the stagnant region and other approaches may have to be considered.

The present analysis is a wafer-scale modeling assuming that the wafer and the pad surfaces are smooth and parallel to one another. However, these surfaces are not quite smooth but has small-scale variations whose characteristic sizes are equivalent to the feature size of a few micrometers. While the presence of these small features influences the flow patterns of the slurry in the gap, it may not be significant and the analysis presented above is mostly valid. If larger scale variations are present, on the other hand, the vertical (or z-directional) motion of the slurry can be induced reducing the size of the stagnant region. This may be achieved by introducing grooves or patterns on the pad surface. For the vertical mixing of the slurry, the characteristic size of the grooves or patterns should be equivalent to the gap size.

Another way to induce the vertical mixing of the slurry is a modification of the slurry. The slurry for the CMP process contains abrasive particles of 0.1 to 1 µm in diameter for planarization. These particles are much smaller than the gap size of the order 50 µm, and do not significantly influence the flow pattern although their presence affects the rheological properties of the fluid. Inclusion of bigger particles of the order of the gap (i.e., 10 to 20 µm in diameter) will induce the z-directional motion of the slurry enhancing the CMP process. These large particles, however, should be made of non-abrasive soft material since they are bigger than the feature size. Otherwise, planarity of the order smaller than a micrometer cannot be achieved.

SUMMARY AND CONCLUSION

The flow of the slurry in a CMP process has been investigated under the assumption that the wafer and the pad are parallel to one another and their surface is smooth. This wafer-scale model has provided the three dimensional flow field of the slurry, the spatial distribution of the local shear rate imposed on the wafer surface, and the streamline patterns which reflect the transport characteristics of the slurry. The results indicate that

(1) the angular velocities of the pad and the wafer should be the same in order to ensure spatially uniform shear rate imposed on the wafer surface
(2) the slurry transport in the gap between the wafer and the pad can be very poor under the prescribed conditions since the stagnant region where the slurry is not replaced by a freshly fed one is quite large.

These results suggest that inducing the vertical (or z-directional) motion of the slurry is preferable to enhance the CMP process. This may be achieved by introducing grooves or patterns on the pad surface or by adding large non-abrasive particles into the slurry. The characteristic size of the patterns on the groove or the large non-abrasive particles should be of the same order of magnitude as the gap size.

ACKNOWLEDGMENT

The authors would like to acknowledge the financial support of the Engineering Research Center (ERC) for Particle Science and Technology at the University of Florida, the National Science Foundation (NSF) Grant No. EEC-94-02989, and the Industrial Partners of the ERC.

REFERENCES

1. F. W. Preston, J. Soc. Glass Technol. **11**, 214 (1927) .
2. W. J. Patrick, W. L. Guthrie, C. L. Standley, and P. M. Schiable, J. Electrochem. Soc. **138**, 1778 (1991).
3. S. R. Runnel and L. M. Eyman, J. Electroche. Soc. **141**, 1698 (1994).
4. D. Wang, J. Lee, K. Holland, T. Bibby, S. Beaudoin and T. Cale, J. Electrochem. Soc. **144**, 1121 (1997).
5. D. Ouma, B. Stine, R. Divecha, D. Boning, J. Chung, I. Ali, G. Shinn and M. Islamraja, in *Proc. 1st Int. Symp. on CMP*, San Antonio, CA, pp. 164-175 (1996).
6. J. Coppeta, C. Rogers, A. Philipossian, and F. Kaufman, in *Proc. 2nd Int. CMP-MIC*, Santa Clara, CA, pp. 307-314 (1997).
7. C. Srinivasa-Murthy, D. Wang, S. P. Beaudoin, T. Bibby, K. Holland, and T. S. Cale, Thin Soilid Films **308-309**, 533 (1997).

INTERFACIAL PRESSURE MEASUREMENTS AT CHEMICAL MECHANICAL POLISHING INTERFACES

LEI SHAN*, STEVEN DANYLUK*, JOSEPH LEVERT**

*George W. Woodruff School of Mechanical Engineering, Georgia Institute of Technology, Atlanta, GA 30332-0405
** Current address is: Advanced Microelectronic Materials, AlliedSignal Inc., 1349 Moffett Park Dr., Sunnyvale, CA 94089

ABSTRACT

We have found that the entrainment of a slurry between a silicon surface and a polyurethane pad will cause the generation of subambient pressure at that interface. These pressures cause the silicon to be further impressed into the pad. We have measured these pressures and this paper reports on the pressure distribution maps over an area beneath a 100mm diameter silicon wafer. The pressures are generally not uniform. The leading 2/3 of the wafer has subambient pressures of the order of 50kPa and the trailing 1/3 of the wafer has positive pressures of approximately 10kPa. The reasons for the subambient pressures is related to the dynamics of the compression of pad asperities, the boundary effects of the silicon edge, the rebound of the asperities, and re-infiltration of the slurry.

INTRODUCTION

There has been considerable discussion in the literature regarding the nature of the entrainment of slurry in the contact between the silicon wafer and the polyurethane pad in chemical mechanical polishing. Some researchers have stressed the hydrodynamics of this contact where the slurry enters the interface and hydrodynamic forces lift the silicon (1,2). Wear occurs by the transfer of energy from the abrasives to the chemically-modified silicon surface. Other researchers have suggested that wear occurs by asperity contact of the polyurethane pad. Abrasives flow through the interface and the mechanical contact of the abrasive trapped at the asperity/silicon interface causes wear. Our work has been focussed in the measurement of the 'slurry film thickness'. We have designed an apparatus to dynamically measure the 'thickness' of the interface, and we have published a number of papers showing how this is done. In the course of our study we have found that the silicon wafer is impressed into the pad beyond what would be expected considering the external loads. We have measured this 'negative' fluid film thickness and we have shown that this effect is consistent with the model of suction by the pad asperities (3,4,5). As a result of this prior work we have seigned and built a fixture to measure the interfacial pressures at the silicon/pad interface. This paper reports on some of these results.

EXPERIMENT

Experiments were conducted using an apparatus shown in Figure 1. The apparatus consists of a table top polishing platen equipped with a dead weight loading system transmitted by linear bearings and a gimbal (spherical) mount. Shear forces are measured from which the coefficient of friction may be obtained. The inset also shows the sample fixture is made of a 100mm diameter steel 'puck' that has 0.4mm diameter holes drilled through the thickness. Steel tubes were brazed on the outside (exit) locations of these holes and the tubes are connected to a pressuring measurement device such as a piezoelectric transducer or monometer. Figure 2 shows

187

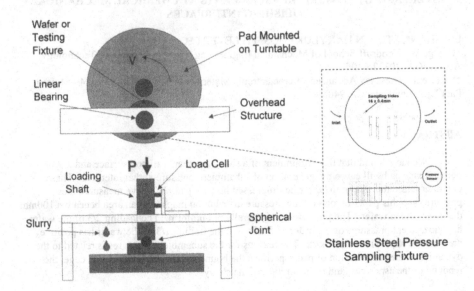

Figure 1. Experimental setup and schematic diagram of both
top and side view of the pressure sampling fixture.

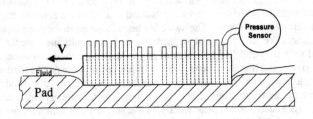

Figure 2. The tubing details of pressure sampling fixture

the details of the pressure tubes in the puck. In the experiments, all tubes are 'pinched off' and the pressure is measured one tube at a time. Pressure maps are obtained by rotating the puck through 45 degrees, recording the pressure and accounting for the differences in linear velocity of the hole locations. The pad used was the Rodel IC1000 and the 'slurry' was water.

The experiments are carried out by conditioning the pad or otherwise modifying the conditions to vary an experimental parameter, starting the pad in rotation, and monitoring the pressure as a function of time from the start of the rotation. A typical result of the pressure (kPa) versus time (min.) plot is shown in Figure 3. This figure shows one example of the pressure measurement at a platen velocity of 0.16m/s, a normal load that produces a average force per unit area of 20kPa. There is a pressure transient with a large decrease to a pressure of 19kPa, then a recovery and a steady state to 10kPa after approximately ten (10) minutes of platen rotation. The results reported will be the steady state values after the ten (10) minute transient.

A typical pressure (kPa) versus distance (hole location) plot for a platen velocity producing a speed of 0.7m/s is shown in Figure 4. Also indicated is the unit normal load 20kPa (3psi), and the average fluid pressure of 19kPa. The data show that the pressure is subambient until the last 10% of the trailing end of the puck. A typical pressure map for a velocity of 0.43m/s is shown in Figure 5. As can be seen, the inlet side of the 'slurry' is subambient and the trailing end is positive. This means that the puck is leaning into the pad and expelling the fluid. Although we have measured this effect over a broad range of experimental conditions, we present examples of four effects, velocity, fluid viscosity, normal unit load, and pad roughness.

Figure 6 shows an example of a pressure (kPa) versus distance (hole location) plot for three velocities 0.16, 0.43 and 0.7 m/s for a unit normal load of 20kPa. The figure shows that the subambient pressure increases, i.e. becomes more negative as the velocity increases. Figure 7 shows the effect of fluid viscosity on the pressure. As the viscosity increases from 1cp (water) to 20cp (8% glycerin), the pressure effect increases. Figure 8 shows the influence of the unit normal load on the pressure. For the two normal loads of 20 and 48kPa the pressure does not change significantly. Figure 9 shows the effect of surface roughness on the pressure. In one case the IC1000 pad was used (roughness of 5x10-6 m) and in the other, the pad was covered with a mylar film that had a roughness of 0.5x 10-6m. Surface roughness has a large effect on the pressure. The pressure is reduced by a factor of three (3) and the trailing edge of the mylar topped pad does not indicate a positive region.

MODEL

We have developed a model for the production of the subambient pressure that includes the asymmetric behavior of the unit normal load, the compression of asperities, and the rebound of the aperities and re-infiltration of the slurry into the valleys of the aperities. Figure 10 shows a summary of this model. The contact stress may be predicted by the contact stress equation shown in the figure (6). This equation predicts that the stress increases at the edges of the wafer and the motion of the wafer skews the stress so that it is higher at the leading edge and smaller at the trailing edge. This stress distribution will compress the aperisities at the edges and these rebound in the interior of the wafer when the stress decreases. The fluid film thickness may be obtained using an exponential asperity height distribution and the Greenwood and Williamson model (7) as shown in the same figure. Finally, Raynolds equation is used to calculate the interfacial fluid pressure (8). These equations predict a converging then a diverging average film profile. This profile results in negative and positive regions beneath the wafer. An example of a

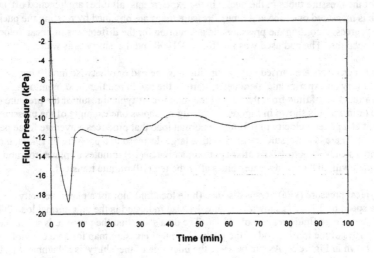

Figure 3. Pressure variation vs. data sampling time, for a pad roughness of 5μm, velocity of 0.16m/sec, and 20kPa normal load.

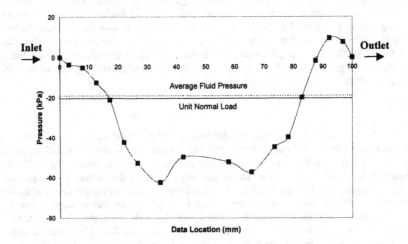

Figure 4. A typical pressure (kPa) vs. distance (hole location) plot for a relative velocity of 0.7m/s and unit normal load of 20kPa.

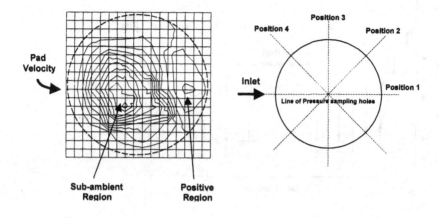

Figure 5. A typical 2-D interfacial fluid pressure map for a velocity of 0.43m/s
and unit normal load of 20kPa.

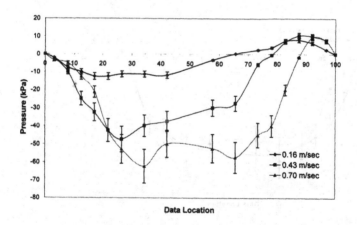

Figure 6. Pressure (kPa) vs. distance (hole location) plot for three velocities
0.16, 0.43 and 0.7 m/s for a unit normal load of 20kPa.

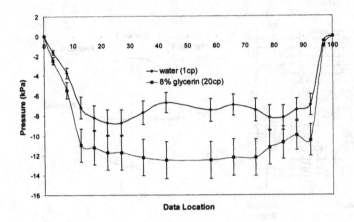

Figure 7. The effect of fluid viscosity on the interfacial fluid pressure
for a velocity of 0.16m/s and unit normal load of 20kPa.

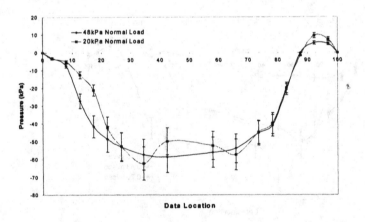

Figure 8. The effect of the unit normal load on the interfacial fluid pressure
for a velocity of 0.7m/s and unit normal load of 20kPa.

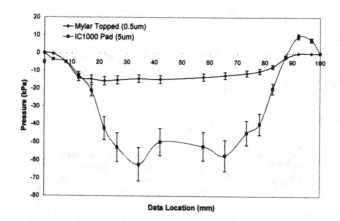

Figure 9. The effect of surface roughness on the interfacial fluid pressure for a velocity of 0.7m/s and unit normal load of 20kPa.

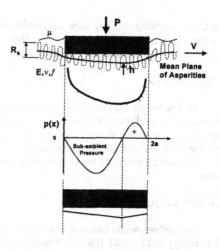

✤ **Contact stress:**

$$\sigma(x) = \frac{P \cos \pi \gamma}{\pi (a^2 - x^2)^{1/2}} \left(\frac{a+x}{a-x} \right)^\gamma$$

✤ **Fluid film thickness:**

$$h(x) = s \ln \left(\frac{1.77 \eta E R^{1/2} s^{3/2}}{(1-v^2)\sigma(x)} \right)$$

✤ **Fluid Pressure:**

$$\frac{d}{dx} \left(h^3 \frac{dp}{dx} \right) = 6 \mu V \frac{dh}{dx}$$

Figure 10. A summary of the analytical model developed for fluid pressure prediction.

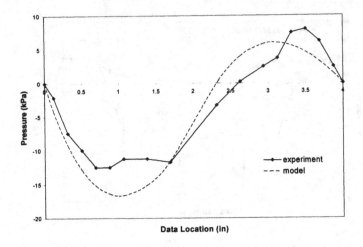

Figure 11. An example of a calculation using the model compared with experimental results for a velocity of 0.43m/s and unit normal load of 20kPa.

calculation compared with experimental results is shown in Figure 11. The model calculations are a fair representation of the data showing that a subambient pressure can be predicted for the first 2/3 of the front of the wafer. A positive pressure results at the rear end of the wafer.

CONCLUSIONS

This paper has summarized some of the results related to the fluid pressure distribution between a hard (steel) surface and a soft compliant (polyurethane) surface. The results are pertinent to the mechanical aspects of chemical mechanical polishing of silicon. The results show that the model can predict a subambient pressure. This is due to the fact that the leading edge of the wafer 'squeegees' the slurry, and the stress distribution beneath the wafer is non-uniform.

ACKNOWLEDGEMENTS

The authors would like to acknowledge the support of the Center for surface Engineering and Tribology and the NSF IUCRC. Special thanks go to Alex Schwartzkopf for this support. Additional thanks go Richard Baker of Rodel for supplying the pads used in these experiments.

REFERENCES

1. S. Runnels, *"Tribology Analysis of Chemical-Mechanical Polishing"*, Journal of the Electrochemical Society, v 141 n 6, 1994, p1698

2. L. M. Cook, J. F. Wang, D. B. James, A. R. Sethuraman, *"Theoretical and Practical Aspects of Dielectric and Metal CMP"*, Semiconductor International, November 1995, p141-144

3. J. Levert, R. Baker, F. Mess, R. Salant and S. Danyluk, *"Mechanisms of Chemical Mechanical Polishing of SiO₂ dielectric on Integrated Circuits"*, to be published in Tribology Transactions, 1998

4. J. Tichy, J. Levert, L. Shan, and S. Danyluk, *"Contact Mechanics and Lubrication Hydrodynamics of Chemical-Mechanical Polishing"*, to be published in Journal of The Electrochemical Society, 1998

5. J. Levert, F. Mess, L. Grote, M. Dmytrychenko, L. Cook, S. Danyluk, *"Slurry Film Thickness Measurements in Float and Semi-Permeable and Permeable Polishing Pad Geometries"*, Proceedings of International Tribology Conference, Yokahama 1995, p1777-1780, Published by the Japanese Society of Tribologists.

6. K. L. Johnson, *"Contact Mechanics"*, P41, Cambridge University Press, Cambridge, UK, 1985

7. Greenwood, J. A. and Williamson, J. B. P., 1966, *"Contact of Nominally Flat Rough Surfaces"*, Proc Roy Soc London, A295, p300-319.

8. Bernard J. Hamrock, *"Fundamentals of Fluid Film Lubrication"*, P141, McGraw-Hill, 1994

PATTERN DEPENDENT MODELING FOR
CMP OPTIMIZATION AND CONTROL

D. BONING, B. LEE, C. OJI, D. OUMA, T. PARK, T. SMITH, and T. TUGBAWA
Massachusetts Institute of Technology, Microsystems Technology Laboratories,
EECS, Room 39-567, Cambridge, MA 02139

ABSTRACT

In previous work, we have formalized the notions of "planarization length" and "planarization response function" as key parameters that characterize a given CMP consumable set and process. Once extracted through experiments using carefully designed characterization mask sets, these parameters can be used to predict polish performance in CMP for arbitrary product layouts. The methodology has proven effective at predicting oxide interlevel dielectric planarization results.

In this work, we discuss extensions of layout pattern dependent CMP modeling. These improvements include integrated up and down area polish modeling; this is needed to account for both density dependent effects, and step height limits or step height perturbations on the density model. Second, we discuss applications of the model to process optimization, process control (e.g. feedback compensation of equipment drifts), and shallow trench isolation (STI) polish. Third, we propose a framework for the modeling of pattern dependent effects in copper CMP. The framework includes "removal rate diagrams" which concisely capture dishing height and step height dependencies in dual material polish processes.

I. MOTIVATION: PATTERN DEPENDENT CMP CONCERNS

The motivation for this work is the presence of substantial pattern dependencies in CMP. As illustrated in Fig. 1, these concerns arise in a variety of key CMP process applications. In oxide or interlevel dielectric (ILD) CMP, the global planarity or oxide thickness differences in different regions across the chip is a key concern. In addition, the remaining local step height (or height differences in the oxide over patterned features and between patterned features) may also be of concern, although such local step heights are typically small compared to the global nonplanarity across the chip resulting from pattern density dependent planarization. In shallow trench isolation (STI), one is typically concerned about dishing within oxide features resulting from over-polish, as well as the erosion of supporting nitride and in some cases the details of the corner rounding near active areas. In metal polishing (such as in copper damascene), one is concerned also with dishing into metal lines, as well as the erosion of supporting oxide or dielectric spaces in arrays between lines.

In this paper, we begin by reviewing previous work on characterization and modeling of oxide CMP pattern dependencies. In Section II, we review the density-dependent oxide CMP model, as well as the important determination of "effective density" based upon a planarization length or planarization response function determination. In Section III, we also review a recent advance in oxide modeling, through which a step height dependent model (proposed elsewhere) has been integrated with the effective density model to produce an integrated time-evolution model for improved accuracy in step height and down area polish prediction. In Section IV we present example applications of the oxide characterization and modeling methodology. These include, first, some comments on the importance of such effects in the design and optimization of the CMP process, and second, an example in which pattern dependent models are integrated with a run by

197

run feedback control scheme to enable pattern dependent oxide control. As a third example, the application of oxide CMP modeling to STI process issues is discussed. In Section V we further generalize the modeling approach to enable application to pattern dependent issues in copper CMP. In particular, we contribute a concise "removal rate diagram" concept that helps to identify the key step height and material dependencies in copper CMP modeling. Finally, a summary is presented in Section VI.

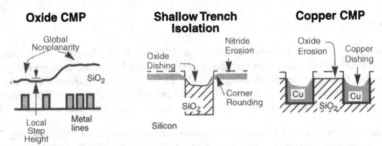

Figure 1. Pattern-dependencies of concern in typical CMP process steps.

II. REVIEW: OXIDE CMP DENSITY DEPENDENT MODELING

In previous work, the fundamental problem we have examined in oxide modeling is the global non-planarity within the die resulting from pattern dependent planarization, as shown in Fig. 2 [1,2]. The goal of our approach has been to achieve efficient chip-level modeling of the oxide thickness across arbitrary product die patterns [3]. The approach has been simplified analytic modeling, in which the removal rate at any location is inversely proportional to the effective density of raised topography. The determination of "effective density" becomes the crucial element in the model: it is through the effective density that the pad and process "sees" that nearby topography influences the polish at any given location.

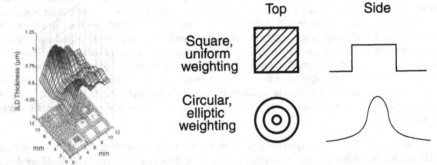

Figure 2. Die-level oxide thickness variation remaining after ILD CMP, resulting from topography arising from pattern dependencies in the underlying metalization.

Figure 3. Window shapes and weighting functions used in calculation of effective density from a given layout.

In our early work, square uniformly weighted windows were used to calculate the effective density, as illustrated in Fig. 3. More recently, we have found substantially better modeling can be achieved with a circularly symmetric, elliptically weighted response function as also shown in Fig. 3 [4]. This response function shape is physically motivated, and corresponds to elastic pad bending or deformation profiles. As part of the methodology for modeling, test chips with "step density" test patterns have been devised to improve the extraction of the planarization response length parameter.The key elements of the density-based model are that the removal rate at any given location depends on the effective density at that location:

$$PR = \frac{K}{\rho(x, y, PL)} \qquad (1)$$

where K is the blanket film polish rate, x and y are spatial locations on the die, and ρ is the effective density which depends on the PL or planarization length (size of the weighting window) used to average local densities. The computation of effective density is illustrated in Fig. 4 for the dielectric characterization mask, which contains regions with gradual changes in local block density, regions with large step density changes between blocks, and regions with equal designed density (50%) and varying pitch. The resulting effective density map is directly related to the final oxide thickness over patterned features observed at the completion of oxide CMP, enabling the extraction of the planarization length from such characterization data.

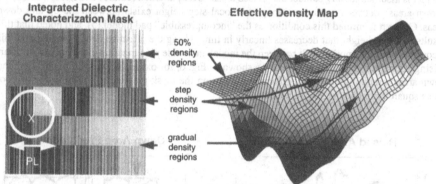

Figure 4. Calculation of effective density for a given layout using the elliptically weighted planarization response function. The characterization mask shown includes structures clearly important in the effective density map [3].

In previous work, details of the planarization response function and its relationship to the pad and process conditions has been discussed [4]. As shown in Fig. 5, the "shape" or weighting of the elliptic response is based on the deformation of elastic material (e.g. the pad) under a spatially localized load of width L. The resulting deformation $w(r)$ can be expressed as:

$$w(r) = \frac{4(1-v^2)qa}{\pi E} \int_0^{\frac{\pi}{2}} \sqrt{1 - \frac{r^2}{a^2}\sin^2\theta}\, d\theta \qquad \text{within the load area } (r<a)$$

$$w(r) = \frac{4(1-v^2)qr}{\pi E} \left[\int_0^{\frac{\pi}{2}} \sqrt{1 - \frac{a^2}{r^2}\sin^2\theta}\, d\theta - \left(1 - \frac{a^2}{r^2}\right) \int_0^{\frac{\pi}{2}} \frac{d\theta}{\sqrt{1 - \frac{a^2}{r^2}\sin^2\theta}} \right] \qquad \text{outside load area } (r>a)$$

(2)

with E being Young's modulus, ν Poisson's ratio, and q the load. Using this density weighting function, it becomes possible to efficiently calculate the effective density for an entire layout as shown in Fig. 4 using a simple 2D convolution (or equivalently, a Fourier transform and inverse transform manipulation) of the weighting function and local layout density.

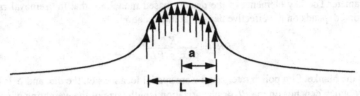

Figure 5. Planarization response function based on elliptic deformation response of an elastic material to a localized load.

Under further simplified assumptions as detailed in [1], excellent fits of the density dependent oxide polish model to experimental data can be achieved. A key assumption in the basic density dependent models is that there exists at all times "planar contact" of the pad on the feature scale: the pad is assumed to only contact the "raised areas" over patterned features and not to contact the "down areas" between features so long as a local step height exists between the up and down areas. Grillaert identified this condition as the "incompressible" pad model [5], and noted that this results in a step height that decreases linearly in time. Using the effective density model and this incompressible assumption, a fit between the model and oxide planarization experiments with surprisingly small error can be achieved. As shown in Fig. 6, for oxide thickness in both raised and down areas remaining after oxide planarization using the mask previously shown in Fig. 4, root mean square errors below 300 Å can be achieved.

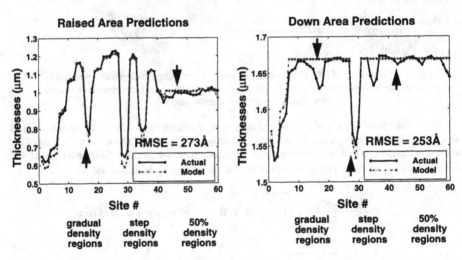

Figure 6. MIT density dependent model applied to both "up" (over metal) and "down" (between metal) oxide thickness resulting from oxide CMP:

III. INTEGRATED MODEL: EFFECTIVE DENSITY AND STEP HEIGHT

An alternative model for the dependence of local step height reduction has been proposed by numerous workers, including Burke [6], and Tseng [7]. In this case, it is proposed that the rate of step height reduction is proportional to the remaining step height. This is based on relative rates of material removal on the up and the down features depending on an allocation of the down pressure between these up and down regions. Under this "compressible pad" model, the step height exponentially decays in time as the local pad pressure differential decreases leading to less effective reduction in time of the smaller and smaller remaining step height. Grillaert et al. [5] proposed a transition model illustrated in Fig. 7 in which large step heights sufficient for the pad to entirely lose contact with the down areas are initially present (during which the planarization proceeds as under the incompressible pad model), and at some transition height h_1 or contact time t_c the pad begins to contact the down areas and the compressible model is then appropriate:

$$h_1 = h_0 - \left(t_c \cdot \frac{K}{\rho} \right)$$

(3)

In that work, a density dependence is also indicated and verified but required or assumed use of the designed local density for large array test structures.

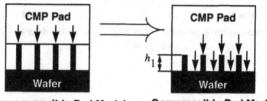

Incompressible Pad Model   Compressible Pad Model

Figure 7. Transition from the incompressible pad model to a compressible pad model as proposed by Grillaert et al. [5]. The transition occurs at some transition height h_1 or contact time t_c.

In examining the regions where substantial errors exist in the simple effective density model of Fig. 6, Smith et al. [8] identified that many of the errors appear to be due to the idealized incompressible pad model. In order to gain the benefits of efficient effective density determination from the MIT model and the benefits of accurate down area and step height determination from the Grillaert et al. model, an integrated approach was proposed. While the original MIT model produced results that depend on two extracted model parameter -- the planarization length (PL) and the blanket removal rate (K) -- the integrated model also extracts the exponential step height decay time constant (τ) and the contact height (h_1) for each site. In the integrated model, the contact height is forced to depend on the effective pattern density as follows:

$$h_1 = a_1 + a_2 \cdot e^{-\rho/a_3}$$

(4)

where a_1, a_2, and a_3 are parameters that are also extracted [8]. The extraction or fit is performed as two iterative loops: the outer loop varies the PL to calculate effective density, at which point the inner loop performs an optimization to find the best values for the other model coefficients that minimize model error. These loops are iterated to find the best set of the six model parameters which minimize the overall squared error.

The resulting model fit is substantially improved, as illustrated in Fig. 8. In particular, the large errors where the effective density model over-estimated the down area polish have been dramatically reduced. The remaining error is on the order of 100Å as compared to the early RMSE of 273A. With 100Å model error, it becomes possible to contemplate using the model in such activities as feedback control where extreme model accuracy is demanded. On the other hand, some regions (notably the large regions with low density which planarized relatively quickly) with substantial model error are still apparent in Fig. 8. It is conjectured that a large-scale pad bending or flexing limit or effect may be present, such as that recently discussed by Grillaert [9].

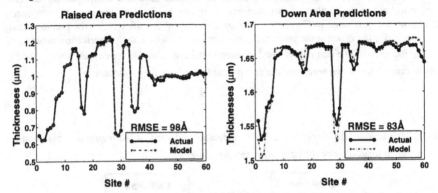

Figure 8. Model predictions and experimental data comparison for the integrated effective density and step height model [8].

IV. CMP APPLICATIONS

In this section, three example applications intimately related to the pattern dependent behavior of CMP are briefly discussed. First, we note the relative importance of die-level effects with respect to typical wafer-scale nonuniformity. Second, we describe recent application of the integrated density and step height model to the run by run control of ILD CMP. Finally, we summarize issues in the application of density models to prediction of shallow trench isolation.

Oxide CMP Process Optimization

The modeling and characterization of pattern dependent variations in oxide CMP further demonstrate the importance of these effects in the design of a viable CMP process. Given the magnitude of the oxide thickness variation present after planarization in comparison to the variation across the wafer as shown in Fig. 9, it is clear that process optimization should be driven by the die-level variation as much as if not more so than by the wafer-level variation.

ILD CMP Run by Run Control

A well-known problem in CMP is that the pad wears over long periods of time as the number of wafers processed on the pad increases. Various strategies have emerged to address this concern, including ex-situ and in-situ pad conditioning, and run by run control. In previous work, we have demonstrated run by run control on blanket wafers, in which polish time and other process parameters are adjusted for the next or later wafer based on measurements on a previous wafer to compensate for removal rate and uniformity degradation [10,11]. In practical use, however, such run

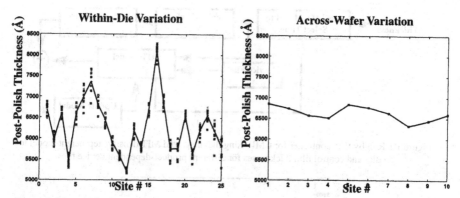

Figure 9. The within-die variation in comparison to across-wafer variation in an ILD CMP process. Shown at left are the thicknesses remaining as a function of 25 site locations within the die; points for five different die across the wafer are indicated. Shown at right is the thickness for the same die position measured at 10 different sites across the wafer.

by run control faces additional barriers to adoption. First, the film thickness measurements have been difficult to acquire on an automated or timely basis. This problem has been largely overcome by the recent availability of the Nova on-line sensor integrated with the CMP tool by which patterned wafer film thickness measurements can be taken on each wafer immediately after polishing [12]. The second barrier to run by run control, however, is the fact that in most fab operations, the same CMP tool and recipe is used across multiple product or product layers. Each product has a unique pattern which, as described above, can have a very large impact on the actual film thicknesses that result within the die on such wafers. As a result, it is typical that retargeting or "look-ahead" wafers are used for each lot of wafers with a different pattern to be processed on the CMP tool.

The ability to accurately model die pattern evolution as discussed in this paper provides a solution applicable to the run by run control of multi-product patterned wafers [13]. As shown in Fig. 10, a feedback control loop incorporating the integrated density and step-height pattern dependent model was developed. For each device type, an appropriate set of model parameters (including effective blanket rate BR and planarization length) were determined. The model for the effective "blanket rate" includes a term $Delta(n)$ that is updated on each run n to track the tool drift in rate over time due to pad and consumable wear:

$$BR(n) = BR(0) + BR\_Device(D) + Delta(n) \qquad (5)$$

The model update is based on seven measurements on four die taken on each patterned wafer using the on-line Nova sensor. The updated blanket rate is then used in conjunction with the pattern dependent model to select the best time for the next wafer of either product type to achieve the desired average target thickness (estimated values from the pattern dependent model of a four die average of 252 sites). The experimental results for a control experiment across two different patterned wafer types is shown in Fig. 11, where we see that +/- 100Å control around the average target thickness has been achieved.

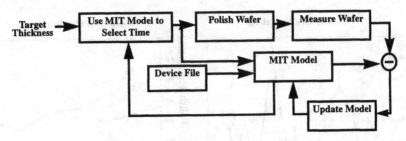

Figure 10. Run by Run controller for CMP using the integrated MIT density/step-height model to predict and control film thicknesses for different product-dependent device files.

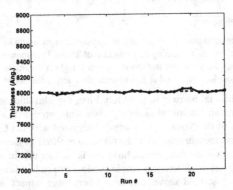

Figure 11. Run by run control over two different pattern wafer types. Wafers were alternated between the two wafer types, with the run by run controller adjusting polish time based on patterned wafer measurements and pattern dependent predictions between each run.

STI Modeling

A third area for application of the effective density and integrated density/step-height models is in the area of shallow trench isolation (STI) formation. As shown in Fig. 1, several pattern dependent concerns arise, including oxide dishing and nitride erosion. In previous work, we have found that good modeling of the oxide overburden stage can be achieved, provided that one takes great care in modeling the effective density resulting from different deposition topographies (e.g. conformal vs. HDP depositions) [14]. We have also found that nitride erosion can be adequately predicted for large active area regions. However, an acceleration of nitride erosion beyond that explain based simply on a density effect is also observed; this issue will be discussed below in the context of copper CMP modeling. Further work to apply the integrated density/step-height model to oxide dishing is also needed.

V. MODEL GENERALIZATION FOR COPPER CMP

To this point we have discussed recent model extensions in the context of oxide CMP. As illustrated schematically in Fig. 1, dishing and erosion concerns in copper damascene processing also require pattern dependent modeling. In this work, we have taken initial steps toward developing a model framework for copper CMP. In particular, we propose a generalization of the density and step-height integrated model for cases where different materials (in this case copper and oxide) are simultaneously undergoing polish. A new notion of "removal rate diagrams" is proposed that concisely captures removal rate dependencies on step height (either as-deposited/plated, or as created due to dishing). We also note additional issues that we believe the model must incorporate to predict copper polish for both wide and fine-line features.

Stages of Copper Polish

A key requirement for copper planarization and polish modeling is the need to track the removal rate of metal and oxide during each stage of the copper polish. In the case of the typical copper damascene process, three different phases or stages in the polish must be addressed, as illustrated in Fig. 12. In the first stage, only copper removal takes place as the copper overburden is planarized and removed. In the second stage, the barrier or liner material must be removed from field and oxide regions; during this stage, it is likely that some degree of dishing in the copper lines may occur. In the third stage, "over-polish" to ensure clearing of the copper and barrier across the entire die and wafer is performed, during which time substantial dishing and erosion may take place in any given structure. In this paper, we focus only on the first and third stages under the assumption that the second stage is very short; similar approaches can also be used to carefully model the second stage.

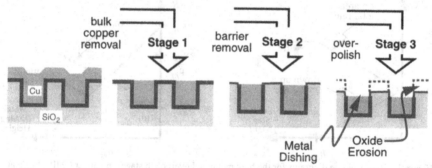

Figure 12. Stages in copper polishing encompassing bulk copper removal, barrier removal, and overpolish creation of dishing and erosion.

"Removal Rate" Diagrams

In the first stage of copper planarization, the integrated density and step-height model described earlier for oxide modeling is again used. A removal rate diagram for this stage is shown at the left of Fig. 13. With time proceeding from the right edge of the local step height (horizontal) axis toward the left, the following occurs. First, a pre-existing step height larger than the "contact" height exists, so that the up copper areas polish at the blanket copper rate modified by the effective pattern density. Note that future work is needed to understand what components or portion of the copper blanket rate are indeed pressure or pattern density dependent; recent work elsewhere on Preston equation modifications [15] are essential to the correct modeling of the impact of density at this and other stages. Continuing in Fig. 13, as time progresses the step height decreases until the contact height h_{ex} is reached; at this point, the copper up area rate decreases linearly with decreasing step height, while the down region rates increase with decreasing step height. When the step height goes to zero local planarity has been achieved and polish continues in both the "up" and "down" areas at the blanket rate.

In the third or overpolish stage, polish rates of both inlaid copper lines or features and of supporting oxide spaces must be considered as shown in the right side of Fig. 13. At the start of this stage, some small (or ideally zero) dishing height is in effect; as time progresses to the right in Fig. 13 this dishing height increases. In the case of "down area" or copper features, the local polish rate decreases linearly with this dishing height; this would continue until some d_{max} height at which pad contact with the down areas is lost and the rate would then go to zero. In the case of the oxide spaces, on the other hand, the local removal rate increases linearly as the dishing height increases. The removal rate diagram clearly indicates an equilibrium or steady-state point (as noted in the case of tungsten polish by Elbel et al. [16]) where both the copper and oxide rates are equal. This steady-state dishing d_{ss} point is the actual maximum degree of dishing one would observe in a given structure.

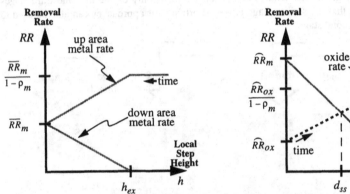

Figure 13. Removal rate diagrams for the bulk metal and overpolish stages in copper CMP. Shown at left is the stage 1 removal rate for the metal up and metal down areas, as a function of the local step height. Shown at right are the stage 2 removal rates for the metal (in trenches) and oxide (spaces) as a function of the dishing height.

Dishing and Oxide Removal Rate Dependencies

The proposed copper model framework described above captures three key effects: step height (or dishing height) dependencies, effective density dependencies, and selectivity between removal in multiple material polish systems. The parameters shown in the removal rate diagrams also need to be extended to account for two additional important pattern dependent effects, as shown in Fig. 14.

First, the d_{max} parameter, which expresses the height at which the pad would lose contact with copper trenches, is also likely to be a pattern dependent parameter. One conjecture is that this parameter may increase as a function of the copper line width, as shown in the left side of Fig. 14.

Alternatively, the d_{max} parameter may be a function of pattern density, as is believed to be the case for contact height in oxide polishing [5,17,8]. More experimental work is needed to identify and characterize this dependency.

A second additional pattern dependent effect needs to be accounted for. As illustrated in the right side of Fig. 14, the effective oxide removal rate (which is modified in the removal rate diagram by the effective density) may also have an oxide space width dependence. That is, the removal rate diagram already accounts for the effective density, but for small oxide space sizes one observes even faster removal of the oxide space than density alone explains. We conjecture a relationship as shown in Fig. 14 based on localized high pressures near the edges of raised features. From contact wear analysis, as the oxide space width becomes small these high pressure peaks extend over a larger portion of the entire oxide feature and accelerate the oxide removal.

d_{max} Depends on Line Width ⠀⠀⠀⠀⠀ RR_{ox} Depends on Line Space

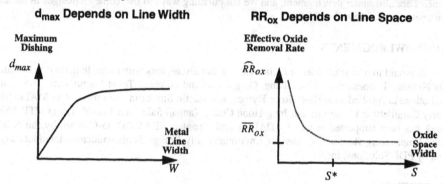

Figure 14. Dishing and oxide erosion rate parameter dependencies. Shown at left is a conjectured dependency between the maximum dishing d_{max} and the metal line width. Shown at right is a schematic acceleration of the oxide removal rate for small oxide space widths.

Copper Model Status: Extraction/Validation of Model

A model framework has been proposed here for prediction of copper polish pattern dependencies. Current work is focused on electrical and physical calibration and validation of model elements. Test masks have been designed which span a variety of pattern parameters, and experimental exploration of process and consumable impact on the polish is in progress [18,19]. The goal is to (a) extract copper model parameters from these experiments, and (b) validate and extend the model as needed for accurate prediction of copper line thickness, copper dishing, and oxide erosion pattern effects. Several challenges need to be addressed. For example, the extraction of planarization length for copper is complicated by the difficulty in making thin film measurements over patterned features in the case of opaque materials such as copper. New approaches, including advanced metrology and alternative mask features may be needed for planarization length and other parameter extraction [20].

VI. SUMMARY AND CONCLUSIONS

In this paper, we have reviewed initial work on oxide planarization modeling encompassing an effective density dependence, and extensions to the model to integrate both effective density and step height dependencies. The integrated model enables efficient full chip oxide thickness prediction with model fits have better than 100Å accuracy achieved. Challenges remain even in oxide modeling, including prediction with especially hard or solo pads, and prediction for alternative slurries and consumable sets.

Several application of oxide models have been presented. These included the importance of within-die characterization and prediction for process optimization, and the use of the pattern dependent model in run by run feedback control. Work has also been done to apply the density model to STI polish. We believe that the generalized framework presented for copper polish is also applicable to the modeling of dishing and erosion in STI CMP.

Finally, we have proposed a model framework for copper CMP simulation. Characterization methods (masks with physical, electrical test structures) for single- and multi-level pattern-dependencies are also under development, and we are pursuing ways to overcome challenges in model parameter extraction and validation.

ACKNOWLEDGMENTS

We would like to acknowledge collaborations and discussions with Dale Hetherington of Sandia National Laboratories; Simon Fang, Greg Shinn and others at Texas Instruments; Tony Pan and others at Applied Materials; Steve Hymes, Konstantin Smekalin and others at SEMATECH; Larry Camilletti at Conexant; and Jung-Hoon Chun, Nannaji Saka, and Jiun-Yu Lai at MIT. This work has been supported in part by DARPA under contact #DABT63-95-C-0088, by the NSF/SRC Engineering Research Center for Environmentally Benign Semiconductor Manufacturing, and by PDF Solutions, Inc.

REFERENCES

1. B. Stine, D. Ouma, R. Divecha, D. Boning, J. Chung, D. Hetherington, I. Ali, G. Shinn, J. Clark O. S. Nakagawa, S.-Y. Oh, "A Closed-Form Analytic Model for ILD Thickness Variation in CMP Processes," *Proc. CMP-MIC Conf.*, Santa Clara, CA, Feb 1997.

2. B. Stine, D. Ouma, R. Divecha, D. Boning, J. Chung, D. Hetherington, C. R. Harwood, O. S. Nakagawa, and S.-Y. Oh, "Rapid Characterization and Modeling of Pattern Dependent Variation in Chemical Mechanical Polishing," *IEEE Trans. on Semi. Manuf.*, vol 11, no. 1 pp. 129-140, Feb 1998.

3. D. Ouma, D. Boning, J. Chung, G. Shinn, L. Olsen, and J. Clark, "An Integrated Characterization and Modeling Methodology for CMP Dielectric Planarization," *International Interconnect Technology Conference*, San Francisco, CA, June 1998.

4. D. Boning, D. Ouma, and J. Chung, "Extraction of Planarization Length and Response Function in Chemical-Mechanical Polishing," *Materials Research Society 1998 Spring Meeting*, San Francisco, CA, May 1998.

5. J. Grillaert, M. Meuris, N. Heylen, K. Devriendt, E. Vrancken, and M. Heyns, "Modelling step height reduction and local removal rates based on pad-substrate interactions," *CMP-MIC*, pp. 79-86, Feb. 1998.

6. P. A. Burke, "Semi-empirical modeling of SiO_2 chemical-mechanical polishing planarization," *Proc. VMIC Conf.*, pp. 379-384, Santa Clara, CA, June 1991.

7. E. Tseng, C. Yi, and H. C. Chen, "A Mechanical Model for DRAM Dielectric Chemical-Mechanical Polishing Process," *CMP-MIC*, pp. 258-265, Santa Clara, CA, Feb. 1997.

8. T. H. Smith, D. Boning, S. J. Fang, G. B. Shinn, and J. A. Stefani, "A CMP Model Combining Density and Time Dependencies," *Proc. CMP-MIC*, Santa Clara, CA, Feb. 1999.

9. J. Grillaert, M. Meuris, E. Vrancken, K. Devriendt, W. Fyen, and M. Heyns, "Modelling the Influence of Pad Bending on the Planarization Performance During CMP," *Materials Research Society 1999 Spring Meeting*, San Francisco, CA, April 1999.

10. T. Smith, D. Boning, J. Moyne, A. Hurwitz, and J. Curry, "Compensating for CMP Pad Wear Using Run by Run Feedback Control," *VLSI Multilevel Interconnect Conference*, pp. 437-440, Santa Clara, CA, June 1996.

11. D. Boning, A. Hurwitz, J. Moyne, W. Moyne, S. Shellman, T. Smith, J. Taylor, and R. Telfeyan, "Run by Run Control of Chemical Mechanical Polishing," *IEEE Trans. on Components, Packaging, and Manufacturing Technology - Part C*, Vol. 19, No. 1, pp. 307-314, Oct. 1996.

12. T. Smith, S. J. Fang, J. A Stefani, G. B. Shinn, D. S. Boning, and S. W. Butler, "On-line Patterned Wafer Thickness Control of Chemical-Mechanical Polishing," submitted to *Journal of Vacuum Science and Technology A*, November 1998.

13. T. Smith, S. Fang, J. Stefani, G. Shinn, D. Boning and S. Butler, "Device Independent Process Control of Chemical-Mechanical Polishing," *Process Control, Diagnostics, and Modeling in Semiconductor Device Manufacturing III*, 195th Electrochemical Society Meeting, Seattle, WA, May 1999.

14. J. T. Pan, D. Ouma, P. Li, D. Boning, F. Redecker, J. Chung, and J. Whitby, "Planarization and Integration of Shallow Trench Isolation," *VLSI Multilevel Interconnect Conference*, Santa Clara, CA, June 1998.

15. S. Ramarajana and S.V. Babu, "Modified Preston Equation For Metal Polishing: Revisited," *Materials Research Society 1999 Spring Meeting*, San Francisco, CA, April 1999.

16. N. Elbel, B. Neureither, B. Ebersberger, and P. Lahnor, "Tungsten Chemical Mechanical Polishing," *J. Electrochem. Soc.*, Vol. 145, No. 5, pp. 1659-1164, May 1998.

17. J. Grillaert, M. Meuris, N. Heylen, K. Devriendt, E. Vrancken, and M. Heyns, "Modelling step height reduction and local removal rates based on pad-substrate interactions," *Materials Research Society 1998 Spring Meeting*, San Francisco, CA, May 1998.

18. T. Park, T. Tugbawa, J. Yoon, D. Boning, J. Chung, R. Muralidhar, S. Hymes, Y. Gotkis, S. Alamgir, R. Walesa, L. Shumway, G. Wu, F. Zhang, R. Kistler, and J. Hawkins, "Pattern and Process Dependencies in Copper Damascene Chemical Mechanical Polishing Processes," *VLSI Multilevel Interconnect Conference*, Santa Clara, CA, June 1998.

19. T. Park, T. Tugbawa, D. Boning, J. Chung, S. Hymes, R. Muralidhar, B. Wilks, K. Smekalin, G. Bersuker, "Electrical Characterization of Copper Chemical Mechanical Polishing," *Proc. CMP-MIC*, Santa Clara, CA, Feb. 1999.

20. S. Hymes, K. Smekalin, T. Brown, H. Yeung, M. Joffe, M. Banet, T. Park, T. Tugbawa, D. Boning, J. Nguyen, T. West, and W. Sands, "Determination of the Planarization Distance for Copper CMP Process," *Materials Research Society 1999 Spring Meeting*, San Francisco, CA, April 1999.

DETERMINATION OF THE PLANARIZATION DISTANCE
FOR COPPER CMP PROCESS

S, HYMES[1], K. SMEKALIN[1,2], T. BROWN[1], H. YEUNG[3], M. JOFFE[3], M. BANET[3], T. PARK[4], T. TUGBAWA[4], D. BONING[4], J. NGUYEN[5], T. WEST[6], W. SANDS[6]

[1]SEMATECH, 2706 Montopolis Drive, Austin, Texas, 78741-6499; [2]National Semiconductor Corp., 2900 Semiconductor Drive, M/S E-100, Santa Clara, CA 95051; [3]Philips Analytical Systems, 12 Michigan Drive, Natick, MA 01760; [4]Microsystems Technology Laboratories, EECS Department, MIT, 60 Vassar Street, Room 39-567B, Cambridge, MA 02139; [5]IPEC Corporation, 4717 East Hilton Ave, Phoenix AZ, 85034; [6]T. West, W. Sands, Thomas West Inc. 470 Mercury Drive, Sunnyvale, CA, 94086.

ABSTRACT

A planarization monitor has been applied to the copper system to investigate pattern dependencies during copper overburden planarization. Conventional profilometry and a noncontact, acousto-optic measurement tool, the Insite 300, are utilized to quantify the planarization performance in terms of the defined step-height-reduction-ratio (SHRR). Illustrative results as a function of slurry, pad type and process conditions are presented. For a typical stiff-pad copper CMP process, we determined the planarization distance to be approximately 2mm.

INTRODUCTION

Copper has emerged as the leading contender for back-end-of-line metallization for advanced integrated circuits. Lack of a viable copper etch process and depth-of-focus limitations of advanced lithography leads to the chemical-mechanical polishing (CMP) of Damascene structures as the preferred method by which copper-based metallization is formed. CMP is the only known feasible method by which copper metallization can be patterned to the requisite, feature size and global planarity.

The primary polish metrics characterizing the topography generated during the CMP of metal Damascene structures are the dishing of wide metal lines, erosion of dense metal arrays and thinning of the dielectric field region. Erosion is a primary issue at the lower levels of metallization where high metal pattern-density, narrow linewidth structures are formed. Dishing is more prominent at the upper metal layers where single structures of greater area (bus lines and bond pad structures) are present. The associated film thickness reductions will influence the circuit performance through line resistance, interlevel and intralevel capacitance variation [1]. Thus, care must be taken to ameliorate such issues, which stem from both plating and polish process limitations.

For the aggressive topographies of next generation integrated circuits, electroplating is a viable fill technique. However, both 'feature-scale recess' of wide structures and 'array-scale recess' can be significant. Figure 1 displays the as-plated array recess as a function of metal pitch for a 50 % metal density array showing over 2000 Å recess in some cases. Such as-plated single feature and array length-scale topographies will directly influence the CMP performance. A strong drive exists to develop plating processes with small as-plated topography on both the individual feature and array dimensions while maintaining desired across wafer uniformity and fill capabilities. The central concern of this paper is to examine the ability of CMP to planarize as-plated copper topography.

Previous work has examined pattern-dependent planarization issues in oxide CMP processes. The polish rate at a specific site on a patterned wafer has been modeled as an appropriately weighted function of the pattern density over a characteristic region defined by the 'planarization length or distance' [2,3]. Such models assumed perfect planarization efficiency and led to linear step height reduction with time relationships. Models which incorporate compressible pads and partial loading of the down regions led to exponential step height reduction with time relationships after the initial linear component [4]. These models, however, neglect changes in effective pressure due to global non-planarity created by topography. An example of this effect occurs during the polishing of next-level metal which incurs preexisting topography as a result of array erosion at the previous level as shown in Figure 2. The resulting array serpentine lines after polish are thicker as a result of initial recess and shielding by the neighboring field region which lowers the effective pressure on the array.

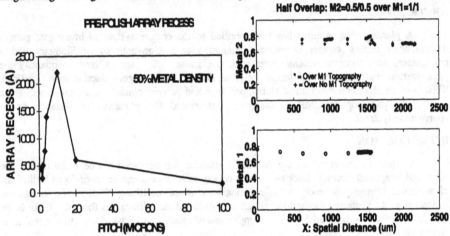

Figure 1 'Array-Scale Recess' resulting from a copper electroplate process over a 50% metal density, 2500μm x 2500μm array as a function of pitch.

Figure 2 Comparison of array metal thickness for an initially recessed array at M2 due to M1 induced topography versus the same array at M2 over M1 field. Field dielectric shielding at M2 reduces M2 array erosion for the initially recessed structures.

In copper CMP processes, two measures are important in characterizing the initial copper removal. The planarization distance indicates the length scale over which raised topography interacts. The planarization efficiency indicates the amount of topography that must be removed to eliminate an existing step height. After the initial copper removal, the CMP process processed with barrier film removal, and during clearing across the wafer also results in erosion and dishing in the multicomponent material system (copper-barrier-dielectric). The resulting within-die total-indicated-range (TIR) of a patterned wafer after CMP is a function of both copper overburden and subsequent dishing and erosion: non-ideal planarization of the as-plated copper topography further contributes to dishing and erosion difficulties.

The initial part of the polish process can thus be viewed as acting to remove preexisting topography while the latter clearing stage aims to prevent the generation of topography. Polish processes having large planarization distances are more effective at both the single feature length scales associated with as-plated line-recess and conventional dishing during the clearing stage as well as the more difficult to planarize array-recess from the plate process and conventional array erosion from barrier clearing during CMP. The planarization efficiency is in general a monotonic function of length scale, worsening with increasing dimension due to pad bending. As the planarization distance of a CMP process increases, the local polish rate becomes increasingly dependent upon the neighboring real estate through the effective pattern density. When the planarization distance becomes much larger than the die size, the effective pattern density becomes a constant and the die is polished the same everywhere (neglecting higher order effects). Harder, stiffer pad processes typically have planarization capability well into the mm-length scale regime, with soft, flexible pads only able to planarize into the 100-micron regime. Hence, in the presence of as-plated array-recess, a hard pad process is desirable.

In this paper, we report the evolution and quantification of the step-height-reduction-ratio, and subsequently extract the planarization distance for copper CMP for the first time. Conventional profilometry and the InSite300™ photo-acoustic measurement tool were employed to quantify copper film thickness and topography. The Insite 300™ operates upon the transient gradient technique and allows for noncontact, nondestructive metal film thickness measurement [5]. The utility of this metal thickness tool bypasses a number of issues which arise with conventional profilometry. The ability to accurately delineate metal feature edge positions and circumvent stress-induced curvature present in long profilometry scans is of principle importance.

EXPERIMENT

The planarization test vehicle is comprised of a set of 18 trenches etched in 200mm dia. Si wafers to the depth of 0.8 μm with the trench set replicated twice per wafer. All trenches were 8.0 mm long, with the width varying from 0.1 mm to 8.0 mm. The 0.1mm structures can be viewed as representing the individual feature performance (dishing of a 100μm wide square) and the higher length scales can be viewed as applying to erosion or array recess effects. The intent was to provide sufficient separation between the trenches in order to isolate each particular structure. However, for the widest structures, the field length was less than 6 mm and the structures begin to interact for processes with a large planarization distance. Additionally, an error in lithography affected some of the larger structures and is discussed later. After silicon etching, a blanket 20 kÅ TEOS layer was deposited followed by 250 Å Ta barrier/1000 Å Cu seed/1.5μm electroplate copper metal film stack. Subsequently, CMP of only copper film (without breakthrough to the underlying barrier or oxide) was performed under a variety on conditions. Films were characterized by either the profilometer or the photo-acoustic tool after each CMP. The resulting step-height after CMP at the middle of the trench relative to the field value at a trench-width distance away was then derived.

RESULTS

Profilometry was performed on a wafer which was polished in 40 second time increments under baseline conditions using an experimental grade copper slurry at 4 psi on a perforated IC1000/Suba IV pad on an IPEC 372MU. Figure 3 displays the extracted step-height-reduction as a function of trench width. Several important conclusions are immediately evident.

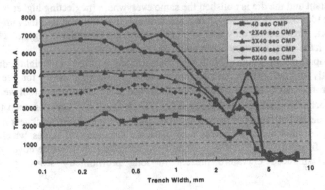

Figure 3 Trench Depth Reduction vs. Trench Width for equal time polish increments

First, no planarization is achieved for feature sizes above 4.5 mm. For feature sizes 1.0 mm to 4.5 mm, only partial planarization is possible, and with the exception of the 3, 3.5 and 4 mm structures, its effectiveness increases monotonically with feature size reduction. And finally, for feature sizes 1.0 mm and below, nearly constant planarization efficiency was achieved, as the trench depth was reduced from 0.8 μm to zero. We believe that the irregularity in our experimental data for 3-4 mm feature sizes was caused by an error in the exposure. This caused each of these structures on wafers from this particular lot to be actually composed of two smaller trenches with slight separation, as evidenced from visual inspection of the pre-polished wafers.

We define the planarization efficiency (PE) in terms of the step-height-reduction–ratio (SHRR), which is the SHR normalized by the amount removed (the maximum SHR value). The planarization distance (PD) is then defined as that length value for which the PE equals a fixed value, e.g. 0.3. This also appears to be the length value at which the inflection point resides. For each polish increment, the inflection point remains on the order of mm as shown in Figure 3, though there does appear to be a slight shifting to lower value. The PD can be viewed as the capability or distance over which the system can planarize, while the PE is a measure as to the speed at which planarization is performed. Based on the data presented in Figure 3, we determined the planarization distance for this copper CMP process to be approximately 2 mm.

Numerous difficulties arose in performing profilometry on the as-polished wafers in some cases. Substrate curvature and feature rounding leading to difficulties in referencing the step edge were two key issues which led to investigation of an alternative metrology approach. Figure 4 displays the associated thickness plots across the full range of structures on the test vehicle using the Insite 300 direct metal thickness measurement tool.

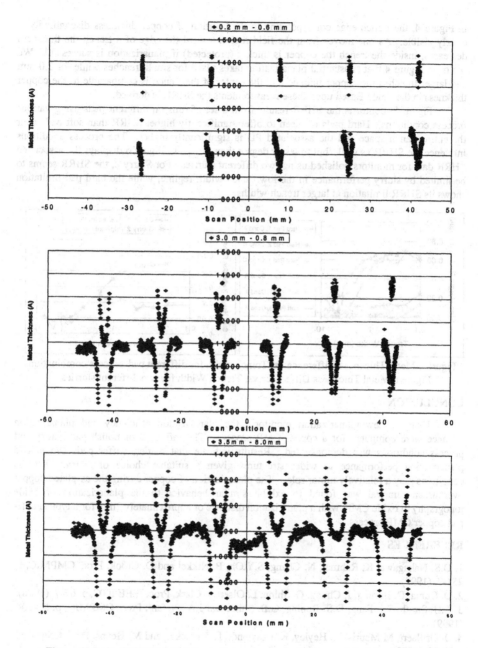

Figure 4. Insite 300 Metal Thickness Scans across the features of the Test Vehicles

In Figure 4, the trench edge corresponding to the position of copper thickness discontinuity is clearly visible. As one moves from the field region toward the edge of a trench the thickness decreases; inside the trench the copper is thicker (protected) if planarization is successful. We see from Figure 4 that substantial planarization takes place for small trenches while for 3.0 mm and larger trenches the copper thickness at the center of the trench is comparable to the copper thickness in the field. Based upon these measurements the SHRR is derived.

Figure 5 contrasts the SHRR results from a set of runs on different pads using various process conditions. Hard pads are seen to offer significantly higher SHRR than soft pads over the entire trench space and the associated PD is significantly higher. The process conditions influence the SHRR as well, but to lesser degree than the pad. Figure 6 displays the associated SHRR data for monitors polished using two different slurries. For Slurry 2, the SHRR seems to be limited by slurry performance in the low trench width regime, while the hard pad limitation shows its SHRR limitation at larger trench widths.

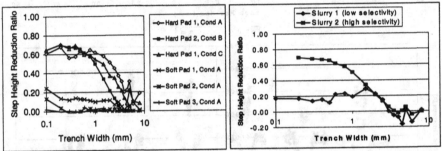

Figure 5 Metal Thickness Difference vs. Trench width for different pad-process combinations
Figure 6 Metal Thickness Difference vs. Trench Width for two different slurries

CONCLUSION

Using a new planarization monitor, the planarization efficiency and planarization distance were computed for a copper CMP process. The influence of polish pad, slurry and process conditions was demonstrated. Results indicate that harder, stiffer pads can extend planarization performance to wider structures given a suitable choice of slurry. Process conditions play a relatively minor role. The planarization of copper during the as-plated copper overburden removal was found to exhibit similar behavior to the planarization of oxide topography in oxide CMP, with a planarization distance of approximately 2mm for a typical stiff-pad copper CMP process.

REFERENCES

1. O.S. Nakagawa, K. Rahmat, N. Chang, S.Y.Oh, P. Nikkel and D. Crook, Proc. CMPMIC, p. 251-7. (1997).
2. D. Ouma, D. Boning, J. Chung, G. Shinn, L. Olsen, J. Clark, Proc. IEEE IITC, p. 67-9, (1998).
3. T.H. Smith, S.J. Fang, D.S. Boning, G.B. Shinn and J.A. Stefani, Proc. CMPMIC, p. 97-104. (1999).
4. J. Grillaert, M Meuris, H. Heyley, K. Devriendt, E. Vrancken and M. Heyns, Proc. CMPMIC, p.79-86, (1998).
5. M.A. Joffe, H. Yeung, M. Fuchs and M. Banet and S. Hymes, Proc. CMPMIC, p. 73-6, (1999).

ON THE PATTERN DEPENDENCY AND SUBSTRATE EFFECTS DURING CHEMICAL-MECHANICAL PLANARIZATION FOR ULSI MANUFACTURING

WEI-TSU TSENG*, JAMES JONG-LIN NIU*, CHI-FA LIN**
*Department of Materials Science and Engineering, National Cheng-Kung University, 1 Ta-Hsueh Road, Tainan 701 TAIWAN. E-mail: wttsen@mail.ncku.edu.tw
**Winbond Electronics Corp., Hsinchu 301, TAIWAN

ABSTRACT

The change of surface profile during chemical-mechanical planarization (CMP) is monitored continuously in this study. The influences from pattern dependency and substrate effects are discussed. Step height reduction rate is a function of pattern density and down force. The rate decreases with time until planarization is achieved. As the polish approaches the patterns underneath, the interaction between substrate effects and pattern dependency results in the resurgence of step height. The implication of this newly found phenomenon is discussed.

INTRODUCTION

The utilization of chemical-mechanical planarization (CMP) process during ULSI manufacturing has been a standard practice in most IC fabs worldwide to planarize the uneven surface topography and to delineate circuit patterns. The popularity of CMP, however, has not been accompanied adequately by a clear understanding of the operating mechanisms involved, in order for a more systematic control of this process. To planarize the wafer surface topography, the polish action has to remove and clear the step height resulting from the deposition process, before the specified planarity can be achieved. This action is complicated by the differential polish rates between the "up" and "down" features, and their dependence on the density of the patterns buried beneath the layer being polished (i.e., pattern dependency).[1-3] Further complexity arises when the polish rates of the same material vary with different substrate materials underneath (i.e., substrate effects).[4,5] Lack of control over the issues described above would lead to loss in planarization efficiency, over or under polish, and poor process reliability.

Some previous modeling and experimental studies have shed some light on the origins of pattern dependency and substrate effects. However, there are contradictions and ambiguities, and further work is still needed before a clear picture can be drawn. Most people attributed the pattern dependency to the uneven partitioning of the down pressure among circuit patterns with different densities and areas, leading to local pressure variation and hence different local removal rate. As a consequence, the step height reduction rate would be a function of local removal rate, which depends on the pattern density. Stine et al. found that the amount of step height reduction is a linear function of time.[1] However, Grillaert et al. considered local pressure resulting from pad deformation and found that, the step height reduction is a linear function of time for an incompressible pad, and an exponential function of time for a compressible pad.[3] Wang et al. reported that the removal rate of blanket oxide wafers decreases or increases depending on the type substrate material beneath (nitride or Al, respectively), when the oxide being polished is thinned down below a certain thickness.[4] To what extent can this substrate effect interact with the pattern dependency to affect the planarization behavior is not clear yet.

In this work, we perform a series of CMP experiments on wafers with different pattern densities and width. The evolution of step height reduction is monitored continuously. The interaction between pattern dependency and substrate effects is observed, and the impacts of these effects on CMP process control are discussed.

Mat. Res. Soc. Symp. Proc. Vol. 566 © 2000 Materials Research Society

EXPERIMENT

Oxide-on-Al and oxide-on-nitride wafers were subjected to CMP experiments. Both blanket and patterned specimens were tested. The silicon nitrides were grown to a thickness of 600 nm on 150 mm Si wafers by plasma-enhanced chemical-vapor deposition (PECVD) with SiH_4 and NH_3 gas chemistry. The Al films were sputter deposited to 600 nm thick, followed by a 40 nm thick TiN layer. PECVD oxides with TEOS (tetraethoxysilane) + N_2O chemistry were deposited on top of the Al and nitride layers to 1600 nm and 2200 nm thick for blanket and patterned wafers, respectively. The Al and nitride layers were patterned using the same test mask, whose layout is shown in Fig. 1. CMP experiments were conducted on an IPEC 372M polisher with IC1000/SubaIV composite pad and silica-based SS-25 slurry. The down force is varied to examine its effects on planarization while the backside pressure is fixed at 3 psi. The table and head speeds were kept at 20 rpm and 25 rpm, respectively throughout the experiments. Oxide thickness was measured by Nanospec and confirmed by SEM. Long-scan surface profiler and AFM were used to characterize the evolution of surface topography during polish.

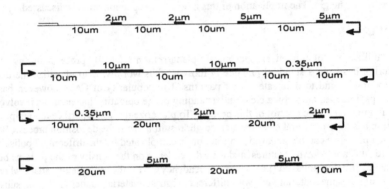

Fig. 1: Pattern layout for Al and nitride layers.

RESULTS

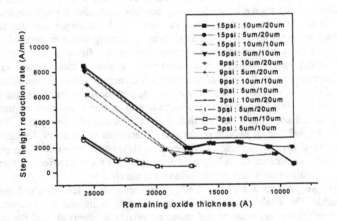

Fig. 2: Step reduction rate vs. remaining oxide thickness.

Initially, the step height reduction rate is very high due to the high local pressure on top of these up features. As time goes by, the up features are reduced rapidly and the pad starts to touch the down features. As a consequence, the pressure is partitioned between the up and down features, causing a fall in the *up* removal rate and a corresponding rise in the *down* removal rate. Finally, planarization is achieved and the removal rate reduced to the blanket removal rate.

Careful examination over the measured data over different patterns reveals that the step height reduction rate is higher for smaller patterns and larger spacings, both of which indicate that local removal rate is higher for lower pattern densities over the range investigated. The results above are similar to previous studies.

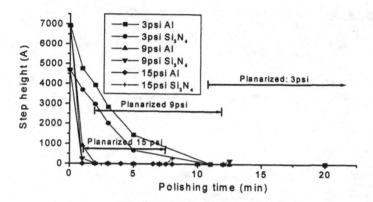

Fig. 3: Changes of step height with time under different down forces.

The change of step height with time is shown in Fig. 3. The Step height is defined as the difference between the up and down features. The time domain within which planarization is achieved are also labeled on the figure. The step height reduces rapidly down to zero and planarization is achieved quickly under a high down force (e.g., 15 psi).

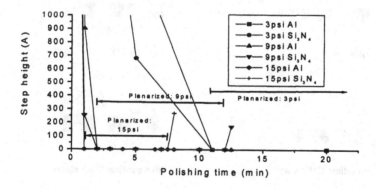

Fig. 4: Changes of step height with time for long polish times.

As the oxide is planarized and thinned down continuously towards the oxide/nitride interface, the removal rate starts to vary again among the nitride patterns. In this case, The removal rates of oxide directly above the nitride patterns are smaller than those in between. As a consequence, the step height emerges again! This is shown in Fig. 4. This unusual behavior is not observed for oxide on Al patterns. The surface profile for oxide on nitride patterns under 15 psi is shown in Fig. 5. The thickness of remaining oxide on top of the nitride patterns is on the order of 25 nm and varies with pattern width and spacing, i.e., it is thicker on top of wider patterns or patterns with shorter spacing. Also noted is that this phenomenon occurs on wafers polished with 15 and 9 psi, but is absent with 3 psi.

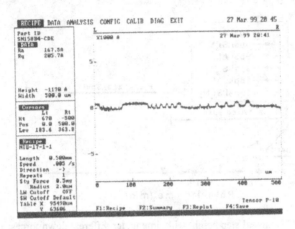

Fig. 5: The resurgence of step height in thin remaining oxide ($\sim$ 25 nm) on nitride patterns after 8 minutes of polishing.

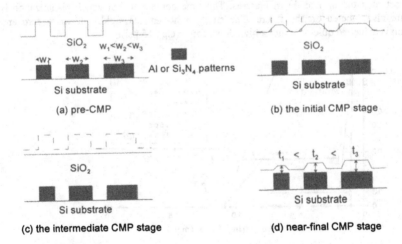

Fig. 6: Evolution of surface profile during CMP. (a) Pre-CMP. (b) Initial stage: step height reduces continuously. (c) Intermediate stage: planarization is achieved. (d) Final stage: Resurgence of step height due to pattern and substrate effects.

The evolution of surface profile during CMP as observed in this study is summarized in Fig. 6. At first, the step height resulting from deposition is quickly reduced as a high local pressure, the magnitude of which depends on pattern density, is exerted on top of the up features. As time goes by, the step height reduction rate decreases as the step height shortens and the down force is partitioned among the up and down features. Subsequently, planarization is achieved and the removal rate approaches that of blanket oxide wafers. Finally, as the polish action proceeds towards as oxide/nitride interface, local removal rate varies again and the step height reappears, whose magnitude depends on pattern density.

DISCUSSION

A few thoughts on pattern dependency:

In this study, pattern dependency during CMP is investigated. It is found that the step height reduction rate is higher for lower pattern densities over the pattern range investigated. While this is similar to various previous studies,[1-3] there are some controversies that should be clarified. If the pattern dependency is indeed caused by the pressure partitioning effects, an isolated pattern (a line or an area) would have to sustain a very high local pressure at the onset of polishing, leading to a very high step height reduction rate. Meanwhile, since this pattern is isolated, the pad would easily touch upon the surrounding "down" areas, and the step height reduction rate decreases quickly, resulting in a low planarization efficiency. On the other hand, for dense patterns, the step height reduction rate would be low to start with, and it may be even lower as the step height decreases since the down features start to get polished. This again would lead to a low planarization efficiency. For an intermediate pattern density, there should appear a maximum planarization efficiency. Experimental and modeling work is underway to pursue this issue.

The scenario above is certainly dependent upon the compressibility of pad and the down force used, as discussed by Grillaert et al.[3] For a perfectly incompressible pad, isolated patterns would be planarized more rapidly. For a perfectly compressible pad, the reverse is true if the *side* polish rate is insignificant. Also worth noting is the effect of table speed, which is not considered in the present study. A high relative velocity has been found effective in improving the planarization efficiency.[1] This may result from the increasing shear rate and hence *side* polish rate. This, in conjunction with a lower down force, would improve the planarization efficiency. This point will be elaborated later.

The resurgence of step height:

The resurgence of step height after planarization is an intriguing phenomenon that is newly found in this study. This may result from the combined effects of pattern and substrate dependencies during CMP. According to Wang et al.,[4] the removal rate (R.R.) depends on the hardness of the substrate material underneath:

$$R.R. = k \left(1/E_f + 1/E_s \right) \times [P/K(w)] \times V \tag{1}$$

where E_f and E_s are the modulus of elasticity for the film being polished and the substrate, respectively; P is pressure, V is velocity, and $K(w)$ is the pattern density factor. Since nitride is harder than oxide, the oxide directly above the nitride patterns *sees* the substrate effect and exhibits a lower local removal rate, while the rest of oxide is still being removed at the same rate. As a result of this local removal rate variation, a step emerges. The observation that this phenomenon is more prominent at wider patterns or shorter spacings seems to suggest that the oxide with harder substrate beneath also experiences a lower down force. Once a step is initiated, the pressure partitioning effects described before is operative again, exacerbating the local removal rate variation and hence a deeper step.

The implications of this step height resurgence phenomenon can not be underlooked. In a sense, the oxide is eroded locally and the step deepens as polish proceeds. This contributes, at least partially, to the dishing effect, which occurs later as the polish gets down to the oxide/nitride interface, where the differential removal rates between oxide and nitride worsens the dishing effects. In other words, dishing is initiated before the oxide/nitride interface is reached. The phenomenon illustrated in Figs. 5 and 6 may resemble that of CMP for shallow trench isolation, where the oxide is being polished down to the nitride polish stop layers. To alleviate this problem, several approaches may be feasible. First, the application of a lower down force will reduce the amount of local the removal rate variation and hence a "shorter" step. In fact, this may account for the observation in Fig. 4 that the step height resurgence is absent under 3 psi, a much lower down force. Second, the introduction of a higher velocity will enhance the shear rate and side erosion rate. This would help remove the steps generated and reduce the amount of dishing. In fact, such an approach has been utilized in STI and metal damascene processes, where the CMP proceeds to the interface. The underlying mechanisms may be due to the scenario proposed above.

The scenario may also account for the dishing effects in metal damascene, where a soft metal is being polished on a hard dielectric. Further work is needed to clarify this point.

CONCLUSION

In this study, we examined evolution of surface profiles during CMP, and the pattern dependency and substrate effects associated with it. Step height reduction rate decreases with polish time or remaining step height, and is also a function of pattern density. After planarization has been achieved, the film is thinned down continuously and step height reappears as the polish approaches the interface between the blanket oxide and the nitride patterns underneath. This phenomenon may result from the interaction between substrate effects and pattern dependency and it may be the origin of dishing effects.

ACKNOWLEDGEMENT

The authors are deeply grateful to Dr. Ming-Shih Tsai of the National Nano Device Laboratories in Hsinchu, Taiwan for his assistance in CMP experiments. This work is supported by the National Science Council of Taiwan under contract number: NSC 88-2216-E-009-015.

REFERENCES

1. B. Stine, D. Ouma, R. Divecha, D. Boning, J. Chung, D. Hetherington, I. Ali, and J. Clark, Proc. CMP-MIC., p. 266 (1997).
2. E. Tseng, C. Yi, and H. C. Chen, Proc. CMP-MIC., p. 258 (1997).
3. J. Grillaert, M. Meuris, N. Heylen, K. Devriendt, E. Vrancken, and M. Heyns, Proc. CMP-MIC., p. 79 (1998).
4. Y.-L. Wang, C. Liu, M.-S. Feng, J. Dun, K.-S. Chou, Thin Solid Films **308-309**, p. 543 (1997).
5. C. G. Kallingal, M. Tomozawa, and S. P. Murarka, J. Electrochem. Soc. **145**, p. 1790 (1998).

STI PATTERN DENSITY CHARACTERIZATION FOR THE SYSTEM ON A CHIP

J. Xie*, K. Rafftesaeth, S.-F. Huang, J. Jensen, R. Nagahara, P. Parimi, B. Ho
LSI Logic Corporation, Santa Clara, CA 95054, *E-mail: JXIE@LSIL.COM

ABSTRACT

Shallow-Trench Isolation (STI) relies on integrated process optimization to achieve the requirement of chip-level process variation across different device features. Characterization of pattern density dependency was investigated through Chemical Mechanical Polishing (CMP) process optimization by Design of Experiment (DOE) and modification of masks by adding dummy structure. Four mask sets with different device features and pattern densities were tested. Effects of slurry selectivity, over-polish extent, silicon trench depth, high-density plasma film thickness, and dummy structure were evaluated. Correlation between physical and electrical data illustrates the process margin in which both logic and memory devices could perform their respective functions. KEYWORDS: Shallow-Trench Isolation, System-on-a-Chip, Pattern Density, Chemical-Mechanical Polishing.

INTRODUCTION

The System-on-a-Chip (SOC) design brings together elements that were previously separate components such as embedded memory technologies, mixed-signal functions and extensive CoreWare® library building blocks. The technology offers customers a single chip to meet multiple market requirements for high-performance products, such as workstations and servers, and low-power, battery-operated devices, including video games and cellular phones etc. [1]. One of the challenges posed on system on a chip is the chemical mechanical polishing for a variety of pattern density blocks. In the shallow-trench isolation process, the high density blocks have relatively lower removal rate and tend to be "under-polished" while the low density ones will lose more oxide and nitride and appears to be "over-polished."

The approach to minimize the chip-level process variation could be achieved through the CMP process optimization and/or made through design/mask modification. Slurry with high oxide-to-nitride selectivity appears attractive for high density devices such as DRAM [2]. While the approach with high selectivity slurry alone may not be adequate for low density devices. It was observed that the erosion loss of nitride and oxide in 20% density feature are significant [3]. By adding dummy nitride structure and/or using reverse etch mask, the erosion could be minimized in the application of conventional slurry [4]. Auto-CAD tool for dummy filling is recently developed and reported being cost-effective [5]. CMP models covering pattern density and polish time dependencies could also be found in recent publications [6-7]. However, the polishing of certain low density features was non-explainable even with the most sophisticated density and time-dependent model [6].

EXPERIMENT

Our process development is conducted on a polishing equipment [8] facilitated with an in-situ laser detection endpoint control. The CMP target (oxide or nitride) is tightly controlled by monitoring a 100μm x 100μm pad structure on the scribeline (called STI monitor). Slurries with oxide-to-nitride selectivity 3:1 and 100:1 (measured on sheet oxide and nitride removal rates)

223

have been evaluated and their respective processes optimized through the Design of Experiments (DOE). Figures 1 and 2 show the trench erosion measured on the STI monitor for both slurries corresponding to CMP DOE process conditions.

Characterization of pattern density dependency are conducted on 4 mask sets with different device blocks and pattern densities:
    Mask A: mixing blocks of high- and low-density devices;
    Mask B: mixing patterns of MIT test structure;
    Mask C: low density blocks only;
    Mask D: low density blocks filled with dummy structure.
Factors evaluated in this program include slurry selectivity, polish time, trench etch, film deposition, and dummy structure. Finally, physical and electrical data are correlated to determine if the process window is robust enough for volume production.

RESULTS AND DISCUSSION

Effect of Slurry Selectivity

Mask A has 4 device blocks of various densities (two of them are identical), as shown in Figure 3. Calculations based on a CAD tool indicates the local chrome densities of the layout ranging from <5% to >50% in 500μm x 500μm cells. Wafers prepared with the mask were polished to certain trench oxide thickness by high-selectivity slurry (HSS) and by low selectivity slurry (LSS) separately. Comparison for device wafers is made based on the difference of the remaining nitride thickness, as shown in Figure 4. On the large feature (STI monitor) and high density structure (test chip), the nitride remaining with HSS is larger than that with LSS; however, on the low density features (memory and logic blocks), the nitride loss with HSS is more significant than that with LSS corresponding to most test conditions of over-polish (10%~40%). Since the low density features are of our most interest, the subsequent discussion will focus on LSS.

Polish Time Dependency

Mask B (MIT test structure) offers a wide range of pattern densities from 0% to 100% by different combination of feature sizes and pitch distances in 2mm x 2mm cells. Wafers prepared with Mask B were polished for 120s, 135s and 150s. The remaining oxide and/or nitride thickness is shown in Figure 5. At 120s, the polish process reaches the nitride layer on the features with pattern densities <50%; but on the features of pattern densities higher than 50%, the thickness of oxide remaining on nitride is up to 1500Å. When the wafers were polished for 135s and 150s, the oxide on the high density area is clean but the loss of nitride on the low density area becomes significant.

Trench Etch and Film Deposition

Trench etch step height and film (oxide/nitride) deposition thickness are two variables for CMP process. Wafers with different silicon trench depth and oxide deposition thickness are polished to approximately the same trench oxide thickness target. As shown in Figure 6, the wafers after polish have similar CMP topography slopes: in a density range of 10~40%, the nitride remaining slope is about 1500Å per 100% density derivative and the oxide remaining on

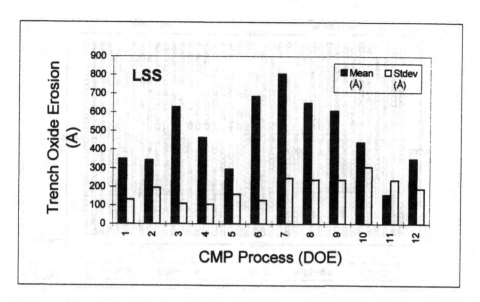

Figure 1: Trench oxide erosion measured on STI monitor (100μm x 100μm) for Low Selectivity Slurry (LSS). Its mean and standard deviation were obtained from 5 measurement sites, corresponding to each CMP process (DOE).

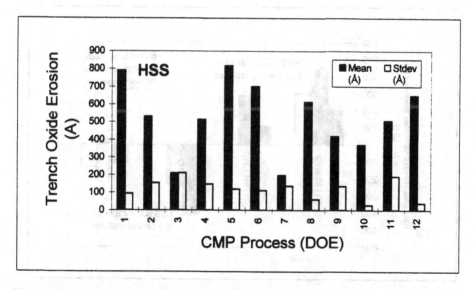

Figure 2: Trench oxide erosion measured on STI monitor (100μm x 100μm) for High Selectivity Slurry (HSS). Its mean and standard deviation were obtained from 5 measurement sites, corresponding to CMP process (DOE).

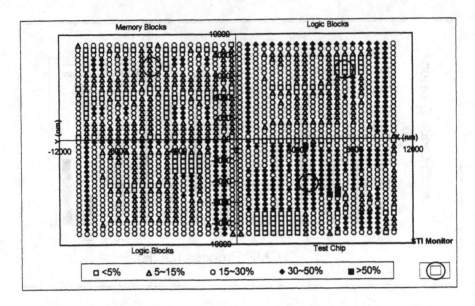

Figure 3: Chrome densities of the layout (Mask A) calculated based on 500μm x 500μm cells. The device includes memory, logic and test structures. Circled area are the CMP monitor sites.

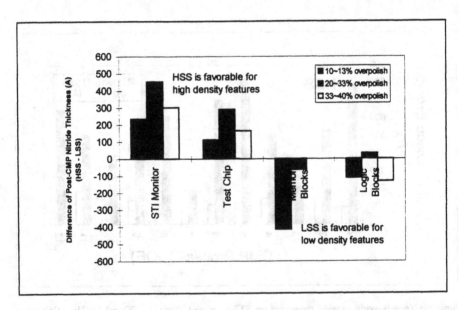

Figure 4: Slurry comparison made by subtraction of the post-CMP remaining nitride thickness (HSS-LSS) corresponding to different density features and over-polish conditions.

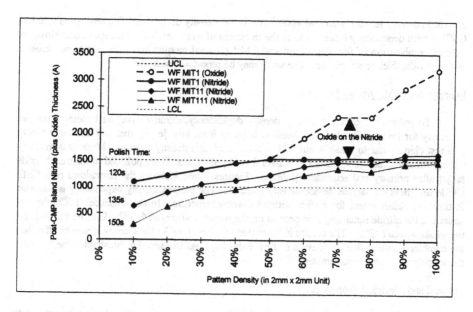

Figure 5: Polish time dependency of post-CMP nitride (plus oxide) thickness measured on different density features using MIT test pattern (Mask B).

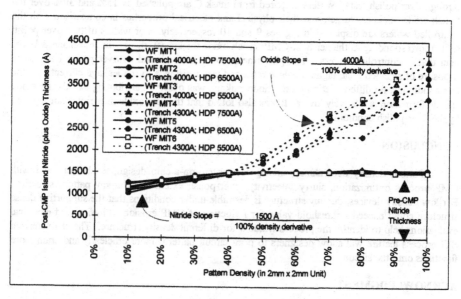

Figure 6: Post-CMP nitride and oxide slopes independent with silicon trench depth and oxide deposition thickness. All the wafers were polished to approximately the same trench oxide target.

50~100% density feature is around 4000Å per 100% density derivative. The results indicate that CMP pattern dependency is unrelated to the thickness of trench etch or film deposition. However optimal combination of film deposition and CMP removal amount will minimize trench erosion and die-to-die thickness variation. The work may be reported elsewhere.

Improvement with Dummy Structure

To reduce the effect of pattern density dependency, dummy structure become attractive especially for low density devices. Mask C is made from low density memory and logic blocks plus test chip similar to Mask A but with reduction of high density features. Mask D is made by adding dummy structure in Mask C. Wafers from Masks C and D are polished to the trench oxide target after primary and secondary polishes. Figures 7 and 8 shows the normalized post-CMP nitride thickness on various device density features. The pattern density dependency is sensitive to the over-polish extent for wafers without dummy structure. In the presence of X% dummy structure, the nitride remaining is improved on the device features of 5~15% densities but little on the weakest area (<5%). The results indicate that in order to make the dummy structure effective, the combined pattern density (device feature + dummy) has to exceed a threshold value to make the supportive structure sustainable.

Physical and Electrical Properties

Pattern density dependency is reflected as the dependency of over-polish extents at which various density features are polished. Their impact on device performance could be examined through over-polish tests. Wafers prepared from mask C are polished by 15% and 30% over the trench oxide thickness target, and their physical and electrical properties in comparison with the controlled wafers are displayed in Figures 9 and 10 respectively. For wafers of two over-polish splits (corresponding to thinner post-CMP trench oxide thickness), the p+ diode leakage is larger than that in controlled wafers. The phenomena could be explained as over polish leads more exposure of silicided area, higher deglaze rates and less resistance to the leakage current. At the test conditions, no differentiation on n+ diode leakage was observed attributing that the n+ diode silicide formation is relatively lower. It was also found that the gate oxide integrity is not affected by the over-polish conditions.

CONCLUSION

Pattern density dependency, arising for system-on-a-chip design, is characterized with CMP process optimization, slurry selectivity, over-polish extent and dummy pattern structure. For low density devices, dummy structure is desirable under conditions that the supportive dense structure could exceed a threshold value for reduction of CMP erosion. Physical and electrical evaluations help to identify the process window for different density structure. The characterized STI process feature will allow designers to print circuit patterns more efficiently and cram more functions onto the silicon.

ACKNOWLEDGMENT

The authors would like to acknowledge their contributions in various phases of this work: Suzette Mack, Ming-Yi Lee, Shumay Dou, Arvind Kamath, Ravindra Kapre, David Heine,

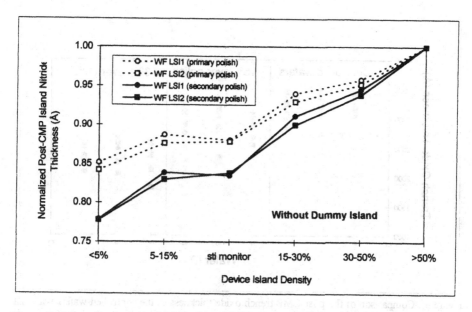

Figure 7: Normalized nitride thickness at different density features for wafers without dummy structure. The pattern density dependency is very sensitive to over-polish sequence.

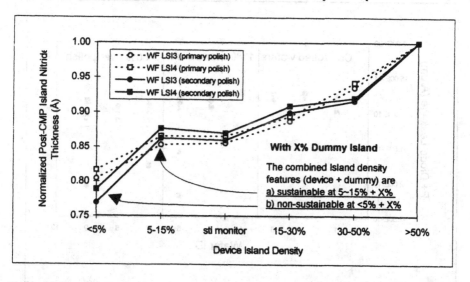

Figure 8: Normalized nitride thickness at different density features for wafers with X% dummy structure. The combined pattern improved certain density features (5~15%) but was not enough for the lowest density area (<5%).

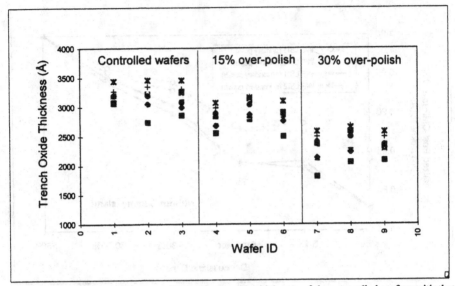

Figure 9: Comparison of the post-CMP trench oxide thickness of the controlled wafers with that of wafers polished for 15% and 30% over the trench oxide thickness target.

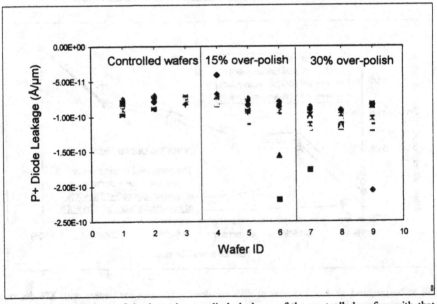

Figure 10: Comparison of the intensive p+ diode leakage of the controlled wafers with that of wafers polished for 15% and 30% over the trench oxide thickness target.

Jayanthi Pallinti, Dawn Lee, Akihisa Ueno and Jim Nunn etc. Reviews from Aldona Butkus, Stanley Yeh and Richard Schinella are greatly appreciated.

REFERENCES

1. S. Pipe, "G12 Technology Sets Standard for System-on-a-Chip Design", *Logically Speaking*, LSI Logic Corporation, Vol. 14 (1998), No.2, p.9

2. J. Grillaert, M. Meuris, E. Vrancken, N. Heylen, K. Devriendt, W.F. Marc Heyns "The Use of a Semi-Empirical CMP Model for the Optimization of the STI Module," *Proceedings of Chemical-Mechanical Planarization for ULSI Multilevel Interconnection Conference (CMP-MIC)*, 1999, pp.105-112

3. T. Smith, S.J. Fang, D.S. Boning, G.B. Shin, J.A. Stefani, "A CMP Model Combining Density and Time Dependencies," *Proceedings of Chemical-Mechanical Planarization for ULSI Multilevel Interconnection Conference (CMP-MIC)*, 1999, pp.97-104

4. G.Y. Liu, R.F. Zhang, K. Hsu, L. Camilletti, "Chip-Level CMP Modeling and Smart Dummy for HDP and Conformal CVD Films," *Proceedings of Chemical-Mechanical Planarization for ULSI Multilevel Interconnection Conference (CMP-MIC)*, 1999, pp.120-127

5. P.-H. Lo, T.-C. Tsai, S. Lin, C-Y. Lee, E. Hsu, H.-C. Wu, H-C Chen, L-M Liu,, "Characterization of Selective-CMP, Dummy Pattern and Reverse Mask Approaches in STI Planarization Process," *Proceedings of Chemical-Mechanical Planarization for ULSI Multilevel Interconnection Conference (CMP-MIC)*, 1999, pp.333-335

6. J. Schlueter, I. Kim, F.J. Krupa, "The Effect of Consumables in the Development of Advanced Shallow Trench Isolation (STI) CMP Processes," *Proceedings of Chemical-Mechanical Planarization for ULSI Multilevel Interconnection Conference (CMP-MIC)*, 1999, pp.336-339

7. E. Tseng, M. Meng, S.C. Peng, "Modeling and Discussions for STI CMP Process," *Proceedings of Chemical-Mechanical Planarization for ULSI Multilevel Interconnection Conference (CMP-MIC)*, 1999, pp.113-119

8. R. Jin, J. David, B. Abbassi, T. Osterheld, F. Redeker, "A Production-Proven Shallow Trench Isolation Solution Using Novel CMP Concepts," *Proceedings of Chemical-Mechanical Planarization for ULSI Multilevel Interconnection Conference (CMP-MIC)*, 1999, pp.314-321

FILM THICKNESS MONITOR FOR CMP PROCESSING

Presenter: Noriyuki Kondo
               Research Lab., Dainippon Screen Mfg. Co., Ltd., Japan

Contributors: Takahisa Hayashi*, Hitoshi Atsuta*, Masahiro Horie*, Masao
               Yoshida**,
               Norio Kimura**, Manabu Tsujimura**, Hiroyuki Yano***, Katsuya
               Okumura***
               *Research Lab., Dainippon Screen Mfg. Co., Ltd., Kamikyo-ku, Kyoto,
               Japan
               **Precision Machinery Group, Ebara Co., Ltd., Fujisawa, 251, Japan
               ***Microelectronics Engineering Lab., Toshiba Co., Ltd., Isogo-
               ku,Yokohama, Japan

ABSTRACT

In the CMP planarization process, detecting the final polishing point is extremely important for maintaining high yield rates.   Therefore, various methods for detecting the final polishing point have been developed.

We consider it vital that the SiO_2 film thickness across a representative area be measured extremely precisely at high speed in the final stages to accurately control film polishing. However, up until now there has not been any wholly satisfactory method to achieve this.

In order to measure a fine representative film area (test pattern) both rapidly and accurately, we concluded that the non-contact, non-destructive and highly reliable methods of spectrometric analysis were most promising.

Based on these considerations, we developed a compact spectrometric film thickness measurement system which can accurately inspect test patterns at high speeds and installed it adjacent to the CMP equipment's polishing table.   Using this system as a film thickness monitor, we can now measure film thickness while rinsing away slurry contaminants, halt polishing at any stage, and calculate the remaining polishing time from the current and desired thicknesses.

This system's pattern identification function can automatically measure the film thickness with very fine test patterns (as small as 40 square microns).   When used with SiO_2 single layer films, the experimental prototype's measurements proved reliably reproducible (showing a deviation of less than 4 nm at 3 σ over repeated trials), and required thirty seconds or less per wafer (including time for pre-alignment and five-point measurement).   The system's theoretical basis and actual operations are described below.

233

CONCEPT OF FINAL POLISHING POINT CONTROL

By accurately measuring the film thickness at the points defined by test patterns just before the completion of wafer polishing, we can precisely calculate the amount still to be polished and thus the remaining polishing time.

To facilitate this analysis, the optical head and measurement unit have been installed next to the polishing table. The measurement unit includes a slurry rinsing function, a pattern-matching auto-alignment function, and a compact alignment stage. Since wafers are measured while held in the top ring of the CMP equipment, polishing can be re-started as soon as a measurement is completed.

The experimental prototype is called the "CMP Semi-InSitu Monitor".

FILM THICKNESS MEASUREMENT

When a sample wafer surface is irradiated with light, multiple reflections occur at the phase boundaries within the thin film, and the spectrum of the reflected light shows their cumulative interference effects. In theory, the spectral composition of the reflected light depends on the film thickness and the wavelength of the initial illumination. Measurements based on these characteristics are commonly called spectrometric, and using this method one can assess film thickness with great ease and accuracy.

Fig. 1 shows the interference effects that occur when a thin film sample is irradiated on a substrate surface. (To clearly demonstrate the phenomenon, the refractive interference is illustrated here with angular incident light.) The reflection of vertically incident light of a specific wavelength depends on the film thickness, which can then be computed if the refractive indices for the intervening medium, film and substrate are all known [1].

Reflectance ratio of single layer

In case of incident angle=0 and absorption=0

$$R = 1 - \frac{8n_0 n_1^2 n_2}{(n_0^2 + n_1^2)(n_1^2 + n_2^2) + 4n_0 n_1^2 n_2 + (n_0^2 - n_1^2)(n_1^2 - n_2^2)\cos 2\delta_1}$$

Notes: $\delta_1 = \frac{2\pi}{\lambda} n_1 d$

no ; refractive index of air
n1 ; refractive index of the thin film
n2 ; refractive index of the substrate
d ; thickness of the thin film

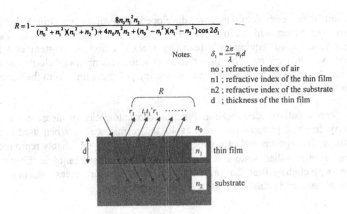

Fig. 1 Light interference on thin film

Fig. 2 shows the theoretical spectral reflectance used to normalize readings for a SiO_2 sample on a Si substrate [2]. The SiO_2 film thicknesses are 500 nm and 600 nm respectively, and we can see that the reflectance varies greatly with small changes in the film thickness. Since conventional analysis methods usually require careful measurement of specific points in the test pattern, the equipment includes a microscope to magnify the pattern points, a spectroscope to measure the spectra of the reflected light, and a data processing section (Fig. 3). While the light source for such testing is generally in the visible range, equipment has been developed which uses UV wavelength light to enhance measurement speed, accuracy and other analytical functions [3]. Finally, since film test patterns should be automatically measured, recent equipment often includes an auto-alignment feature consisting of a two-dimensional CCD camera and an image processing section [4] (Photo. 1).

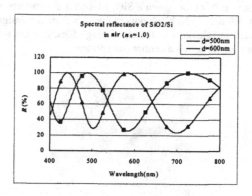

Fig. 2 Theoretical values of spectral reflectance

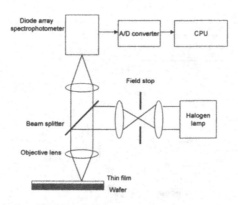

Fig. 3 Block diagram of ordinary spectrometric film thickness measurement system

In previous spectrometric film thickness measurement systems, the spectral values for each of the sample's reference points are calculated individually with a spectroscope. The CMP Semi-InSitu Monitor, however, employs a variable wavelength light source and a new "effective simultaneous measurement" method to gauge film thickness. This method not only measures discrete points on the sample, it measures the film thickness at a number of points at the same time around a target pattern utilizing the same two-dimensional CCD camera used for pattern-matching wafer alignment (Fig. 4). Fig. 5-a shows the actual images obtained with this method and the great differences in image contrast due to interference phenomena. These variations can be elicited simply by changing the wavelength of the incident light while using exactly the same image pattern. This offers a new source of information from which film thickness and other characteristics may be measured as well. Fig. 5-b shows spectral reflectance values from specific test pattern pixels in the above mentioned measured image. Using these reflectance values, we can readily compute the actual film thickness. In this case, given a SiO_2 film on a Si substrate, we can calculate that the film thickness is 900 nm. The black dots in Fig. 5-b show the designated pixel's spectral reflectance of different wavelengths in the eight image files, and the solid line indicates the theoretical reflectance values across a continuous spectrum.

Photo. 1 Configuration of full automatic film thickness measurement system (VM-3000 from SCREEN)

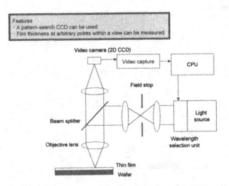

Fig. 4 Block diagram of film thickness measurement section in CMP Semi-InSitu Monitor

236

Since this method captures film thickness information from various areas covered by the CCD camera, the thickness distribution of an area can be measured at extremely high speed. For example, Fig. 6-a shows a three-dimensional plot of the thickness distribution around the Fig. 5-a test pattern taken with the CMP Semi-InSitu Monitor. Please compare this with Fig. 6-b which shows a three-dimensional plot of an 8-inch wafer film measurement taken with another simultaneous film thickness measurement system with different optical specifications.

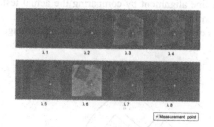

Fig. 5-a Sample test pattern images generated with various different wavelengths

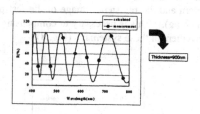

Fig. 5-b Spectral reference values for a specific pixel in the test pattern

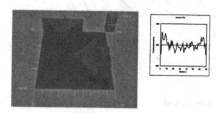

Fig. 6-a Three-dimensional film thickness distribution around the test pattern

Fig. 6-b Three-dimensional film thickness distribution on the 8-inch wafer

WAFER ALIGNMEMT USING AN IMAGE PROCESSING (PATTERN MACHING) METHOD

To measure the film thickness in the specified area (test pattern) on a wafer, precise image pattern comparisons are required on the wafer surface and the optical head must be aligned exactly with the test pattern.

Wafers are ordinarily spinning during CMP processing, and if they are not subsequently aligned, their angular orientations (notch positions) will differ randomly when each is stopped and held by the top ring for measurement. Also, each wafer may deviate slightly from the center of the top ring due to gaps between the wafer edge and the guide ring. Before measuring the film thickness of a specified area on a wafer, therefore, the wafer's position and orientation relative to the optical head must be detected for pre-alignment. Conventional stand-alone film thickness measurement systems are provided with a dedicated sensor for pre-alignment.

The sensor detects the wafer notch while it is spinning in the system's measurement unit so that it can be positioned at the specified angle. However, to simplify operation and miniaturize size, we designed CMP Semi-InSitu Monitor so it would not require a dedicated pre-alignment sensor. This system can detect both center deviation and the wafer notch angle relative to the reference point simply by analyzing the images captured by the same CCD camera used for fine alignment and film thickness measurement. (Fig. 7)

To precisely position the optical head for each test pattern, almost all existing automatic film thickness measurement systems need to perform fine alignment by comparing the actual test pattern and the registered image of the test pattern including their surrounding areas (pattern matching). This is necessary to compensate for deviations during pre-alignment and/or stage movement in these systems. (Fig. 8)

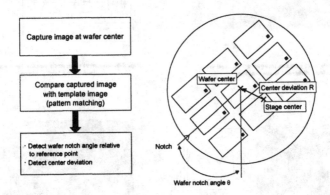

Fig.7 Pre-Alignment

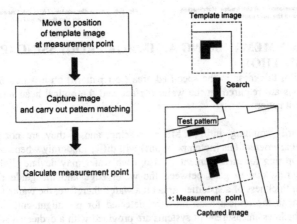

Fig.8 Fine-alignment

However, since the CMP Semi-InSitu Monitor can correctly calculate wafer orientation and center deviation during pre-alignment, its stage can move the optical head almost directly to the required measurement location. You can, of course, also use exact pattern recognition to ensure the most precise possible alignment. To do this, simply register a unique pattern near the measurement target when you create a recipe, and perform pattern matching as the last step in the alignment sequence. However, the general accuracy of this pre-alignment method allows you to measure film thickness even if pattern recognition cannot be performed in certain target areas because there are no unique patterns within the camera's field of view.

After this simple positioning, the head starts to measure the film thickness at the specified point in the target pattern. At this stage, you can also sample several other points at the same time (effective simultaneous measurement) thanks to the two-dimensional CCD field. Specifying additional inspection points not only enhances the performance of your film thickness readings, it also lets you collect a variety of other film information which you can later use for other analytic purposes.

SYSTEM CONFIGURATION

The measurement unit is composed of an optical head to measure film thickness and an R-θ stage to position the optical head below each test pattern on the wafer. To facilitate quick, convenient use, the measurement unit is located adjacent to the polishing table in the CMP equipment. Other components include a control unit, a fluid circulation unit, and an operation unit and they are all located near the CMP equipment. (Fig. 9)

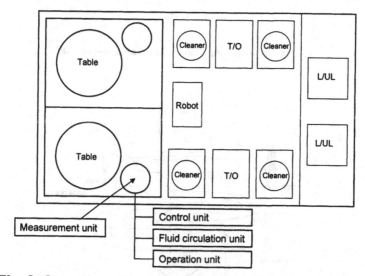

Fig. 9 Overall configuration diagram of CMP Semi-InSitu Monitor

239

MEASUREMENT UNIT

To reduce the size of the measurement unit so that it can be installed in a small space adjacent to the CMP polishing table, we developed a small optical head and used a compact R-θ stage containing both linear and rotation sections to move the head across the wafer. (Figs. 10 and 11)

Slurry can be rinsed out of the measurement area by DI water dispensed from nozzles around the optical head's window, so that a clear image can be input. (Fig. 12) The gap between the wafer surface and measurement head is completely filled with DI water to prevent image degradation caused by water surface effects or air bubbles.

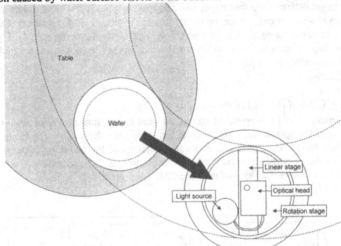

Fig. 10 Polishing table and measurement unit

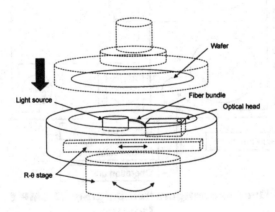

Fig.11 Measurement Unit

240

The configuration of the optical head is illustrated in Fig. 13. The variable wavelength light source emits light which enters the optical head through the fiber bundle. The light travels through the lens, mirror, objective lens, prism, and then reaches the wafer surface through the slurry rinse DI water. The light is reflected from the wafer, returning along the same path, and passes through the beam splitter and transmits the wafer test pattern image to the two-dimensional CCD camera.

Focus can be controlled by monitoring the image contrast with the two-dimensional CCD camera while using the AF actuator to move the objective lens back and forth along the optical axis. Very clear images can be obtained with this simple focus control.

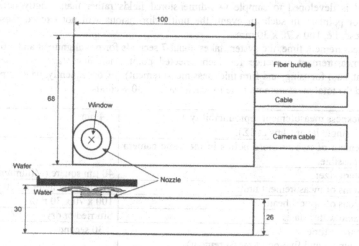

Fig. 12 Optical head and rinse nozzle

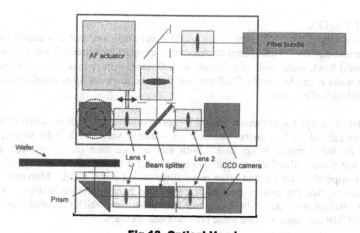

Fig.13 Optical Head

SPECIFICATION AND EVALUATED PERFORMANCE

Table 1 shows the CMP Semi-InSitu Monitor's specifications and the results of a recent performance evaluation. Compared with conventional optical spectrometric film thickness measurement systems, the reproducibility of CMP Semi-InSitu Monitor measurements is slightly lower due to the presence of water on the wafer. Nevertheless, when measuring SiO_2 single films at 3 Σ over repeated trials, the reproducible accuracy is within 4 nm, which is more than sufficient to effectively control CMP processing.

The measurement unit is 310 mm in diameter and 160 mm high, mostly due to the size of the R-θ stage. Although this unit is already quite compact, its size can be even further reduced if a method is developed to sample two-dimensional fields rather than strictly defined test patterns or points. In such an event, the unit's dimensions will not exceed those of the optical head, i.e, 100 x 70 x 30 mm.

As for measurement time, the system takes about 7 seconds for pre-alignment and 4.6 seconds for the measurement sequence for each selected point (including stage movement, fine-alignment, auto focusing, and film thickness measurement). Consequently, when 5 points are measured the total measurement time for each wafer is 30 seconds.

Film thickness measurement reproducibility (for SiO_2 single layer films at 3Σ):	< 4 nm
Measurement of two or more points in the same camera view is possible.	
Test pattern size:	40 μm square (Minimum)
Dimensions of measurement unit:	310 mm diameter x 160 mm
Dimensions of optical head:	100 x 70 x 30 mm
Measurable wafer states:	Slurried or dry
Measurement time (Pre-alignment and five-point measurement):	< 30 seconds

Table 1

CONCLUSION

We have developed the CMP Semi-InSitu Monitor and confirmed that it is capable of highly precise measurements. This system incorporates a compact measurement unit composed of a small optical head, wafer rinse nozzles and an R-θ stage. Since measurement is performed while the wafer is still held in the CMP top ring, it can be immediately returned to the table for further polishing if necessary.

This system employs a simultaneous film thickness measurement method, which incorporates a two-dimensional CCD camera "detector", a variable wavelength light source, and an analyzer for the captured image data. With this configuration the system can, not only measure test pattern film thickness but also be used for a variety of visual wafer checks and film data inspections for Cu and other metal films during metal CMP. Moreover, we now have evidence that the system may even be able to handle moving wafers. With such potential, this system could evolve into a true In-Situ Monitor which measures film thickness inside the CMP unit itself with the same precision and accuracy.

REFERENCES

[1] H.Kubota, "Ouyou-Kougaku-Gairon", p217-218, Kinbara Press.Inc., 1965
[2] E.D.Palik, "Handbook of optical Constants of Solids", pp.547-569, Academic Press.Inc., 1985
[3] N.kondo,et,al., "Film thickness measurement system of ultrathin film using light of UV wavelength", pp.392-402, SPIE Vol.1673, 1992
[4] N.Kondo, "Film thickness measurement using light", OPTRONICS'97, pp125-127, 1997
[4] Extended Abstracts; The Japan Society of Applied Physics.1989:815 pages N.Kondo et al.

REFERENCES

Part VI

Particle Adhesion and Post-Polish Cleaning

POST-CMP MEGASONIC CLEANING USING
DILUTE SC1 SOLUTION

A. A. Busnaina, N. Moumen, J. Piboontum and M. Guarrera.

Microcontamination Research Laboratory
Center for Advanced Materials processing
Clarkson University, Potsdam, NY 13699-5725
http://www.clarkson.edu/~microlab

1. INTRODUCTION

Non contact cleaning or wet-cleaning processes, were the megasonic play a key role in the separation of the particles from the wafer is a commonly used technique in semiconductor manufacturing. CMP process can be very damaging to the production yield if not followed by an effective post clean process. McQueen[1,2] identified the effect of the acoustic boundary layer and its role in the removal of small particles at high frequency. Busnaina et al[3-6] studied ultrasonic and megasonic particle removal and the effect of acoustic streaming. They showed that the cleaning efficiency increased with power until a certain range and then decrease slightly[7,8]. Busnaina et al result indicted that SC1 removes more particles than DI water particularly at lower megasonic powers specially in the case were the slurry particles are deposited onto the wafer surface by dipping experiments. But they also demonstrated that it was possible to achieve 100 % removal in DI water when using the optimum conditions. This paper presents the latest results of the post-CMP megasonic cleaning process, this study is focused on the cleaning of thermal oxide silicone wafer polished using silica based slurry and cleaned using diluted SC1 ($H_2O/H_2O_2/NH_4OH$: 40/2/1).

1.1 Acoustic Streaming

There are three types of acoustic streaming that occur in a sound field. Rayleigh streaming occurs outside the acoustic boundary layer due to a standing wave in a tube or channel. The vortices in Rayleigh streaming are of the scale of the acoustic wavelength. Eckart streaming occurs in a free nonuniform sound field. Boundary layer streaming occurring due to interactions with obstacles in an acoustic flow is referred to as Schlichting streaming. Streaming velocity increases with the increase of frequency and power. The most important aspects of megasonic cleaning is the acoustic boundary layer which is very small compared with a typical hydrodynamic boundary layer at the same velocity. This exposes particles on the surface to a much larger velocities and improves the cleaning efficiency.

1.2 Particle Removal

Particles removal in megasonic cleaning relies on the reduction of the boundary layer thickness on the substrate and the three types of acoustic streaming in the tank. In addition to the megasonic effect in removing contamination from the wafer surface, the use of basic chemistry has shown big improvements in cleaning efficiency. The cleaning mechanism in SC1 ($H_2O/H_2O_2/NH_4OH$) is based on separation of the particles from the substrate, by the introduction of an electrostatic repulsion between particles and wafer. The adsorption of the OH⁻ species on the particles (SiO_2) and wafer, in alkaline pH is responsible for the creation of a negative Zeta potential. The consequence of the electrostatic repulsion is to weaken the total adhesion force between the particles and the wafer and minimizes particle redeposition[9].

Mat. Res. Soc. Symp. Proc. Vol. 566 © 2000 Materials Research Society

2. EXPERIMENTAL

2.1 Experimental Facilities

All the cleaning experiments were done in the Microcontamination Research Laboratory's class 10 clean room at Clarkson University. The equipment used consists of a quartz megasonic overflow tank made by PCT.Inc, with a maximum power of 640 Watts and a frequency of 760 kHz. The polishing experiments were conducted at **RPI** (Troy, NY) using IPEC polisher model 372M. A spin rinser/dryer, by SemiTool Inc. (STI), was used to dry the wafers at the end of the cleaning process. A laser surface scanner made by Particles Measuring Systems, Inc. was used to determine particles counts. The scanner has a capability to detect particle sizes between 0.1 μm to 10 μm.

2.2 Experimental Procedure

The polishing conditions were:
- STI 30N50PHN silica based slurries was used
- Slurry flow rate = 125 ml/min
- Platen speed = 28 rpm
- Platen Temperature = 78 F
- Polishing pressure = 8 PSI
- Polishing time = 122 sec

The wafers were kept wet until cleaning. The SC1 volume ratio used was ($H_2O/H_2O_2/NH_4OH$: 40/2/1). The effect of megasonic power, cleaning time and SC1 temperature on the cleaning efficiency is studied. The experimental conditions for these key variables were determined by the use of statistical design software *(JMP by SAS)*. After cleaning, the wafers were dried using the SemiTool Inc (STI) spin dryer. The wafers were then scanned using the surface scanner and the post particle count was obtained.

3. RESULTS

The results are shown in terms of the number of defects (larger than 0.2 micron) on the wafers in figures 1, 2, 3, 4, 5 and 5. Figure 1 and 2 show the total number of defects above 0.2 μm remaining on the wafer surface, for a fixed cleaning time 10 and 18 minutes at different values of temperature and megasonic power. After cleaning for 10 min, a high number of defects is observed when the megasonic power is between 300W–500W and at low temperature. On the other hand, a lower number of defects is achieved at high temperature and high or low power. For 18 minutes cleaning time (figure 2) a low number of defects is achieved only at high temperature and high power. Figure 3 and 4 presents the number of defects for two different temperatures 30^0C and 68^0C as a function of the megasonic power and time. A low number of defects (33 defects) is achieved at 68^0C above 10 min at high power and at low power for a cleaning time less than 15 minutes. Figure 5 and 6 shows the effect of megasonic power on the removal efficiency. A low number of defects can be achieved at a megasonic power of 48 Watts, and a shorter cleaning time (less than 15 min), and a high temperature (more than 55^oC). At the high power 640 watts (maximum power) the minimum defects is achieved only at high temperature and longer time

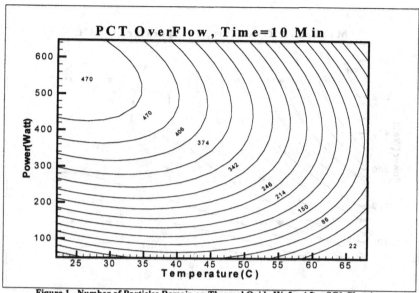

Figure 1. Number of Particles Remain on Thermal Oxide Wafer After SC1 Cleaning

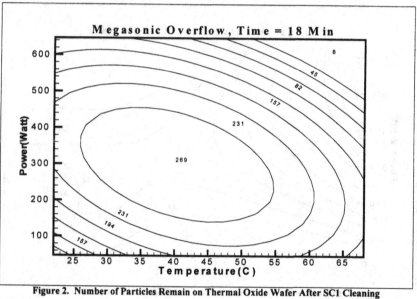

Figure 2. Number of Particles Remain on Thermal Oxide Wafer After SC1 Cleaning

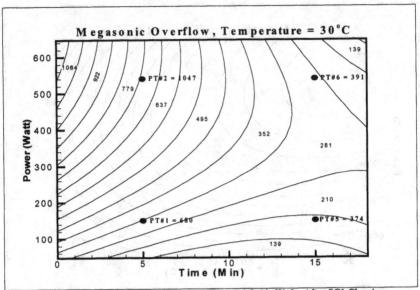

Figure 3. Number of Particles Remain on Thermal Oxide Wafer After SC1 Cleaning

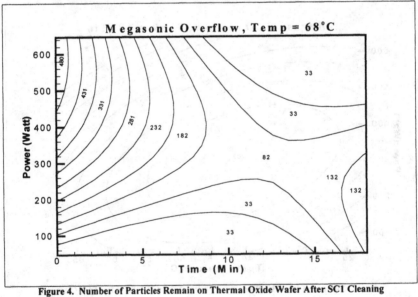

Figure 4. Number of Particles Remain on Thermal Oxide Wafer After SC1 Cleaning

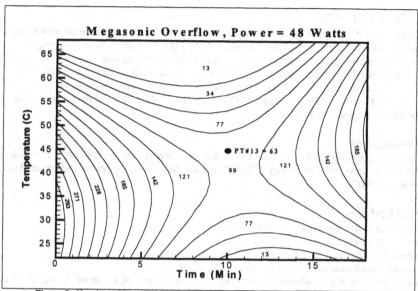

Figure 5. Number of Particles Remain on Thermal Oxide Wafer After SC1 Cleaning

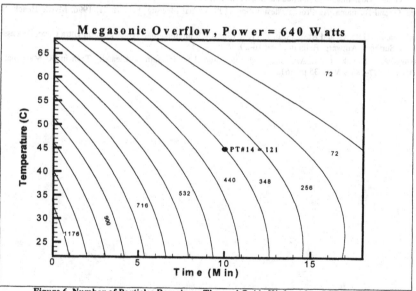

Figure 6. Number of Particles Remain on Thermal Oxide Wafer After SC1 Cleaning

4. CONCLUSIONS

This study shows the effect of the three key parameters (power, time and temperature) on a post-CMP non-contact cleaning process using a diluted SC1. Effective particle removal using megasonics with dilute SC1 chemistry is achieved from thermal oxide wafers polished using STI silica slurry (down to 38 defects per wafer). The relation between those cleaning parameters is often complicated by other defects (scratches, etc.) which is counted as particulate contamination. A diluted SC1 as the one used in this study ($H_2O/H_2O_2/NH_4OH$: 40/2/1) has shown a very good removal efficiency and no risk of roughness.

Acknowledgment

The authors gratefully acknowledge the financial support by SRC-Center for Advanced Interconnect Science and Technology (CAIST) and the New York Center for Advanced Materials Processing (CAMP).

REFERENCES

(1) McQueen, D. H., Ultrasonics 24, 273 (1986).

(2) McQueen, D. H., Ultrasonics 28, 422 (1990).

(3) Kashkoush, I. and Busnaina, A., J. Particulate Sci. Tech. 11, (1993)

(4) Busnaina, A and Kashkoush, I., Chem. Eng. Comm. 125, 47, (1993).

(5) Gale, G., Busnaina, A and Kashkoush, I.,Proceedings, Precision Cleaning 94, Rosemont, IL, May 17-19, pp 232 (1994).

(6) Gale, G. and Busnaina, A., J. Particulate Sci. Tech. 13,197 (1993)

(7) Gale, G. and Busnaina, A., Proc. Adhesion Soc., 19th Annual Meeting, Feb. 18-21, 1996, Myrtle Beach, SC, pp 520.

(8) Busnaina, A., Gale, G. and Kashkoush, I., Proc. 13th ICCS International Symposuim on Ultra Clean Processing of Silicon Surfaces, Antwerp, Belgium, Sept 16-20, 1996.

(9) Kashkoush, I. Novak. R. and Al., Proc. Fifth International Symposium on Cleaning Tech in SC Manufacturing., ISBN 1-56677-188-9 V 97-35 pp 161.

MECHANISM OF A NEW POST CMP CLEANING FOR TRENCH ISOLATION PROCESS

N.Miyashita*, Y.Mase*, J.Takayasu*, Y.Minami*, M.Kodera*, M.Abe*, T.Izumi**
* Microelectronics Center, Toshiba Corporation
8, Shinsugita-cho, Isogo-ku, Yokohama 235-8522, Kanagawa Japan
** Department of electronics, Tokai Univ.
1117 kitakaname, hiratsuka, Kanagawa, Japan

ABSTRACT

CMP has been revealed as an attractive technique to poly Si of trench planalizing process. Major issues of process integration for that purpose have been post-CMP cleaning process. A new post CMP cleaning process which employed special organic surfactant has been reported in this paper. In general, wafers after CMP process are contaminated by particles and metallic impurities in the case of conventional cleaning method. The contamination introduce the defects into the wafers after oxidation. The contamination was removed by new cleaning method, using DI water containing special organic surfactant and silica particles. The experimental work has focused on critical problems that had to be solved, using AFM, EDX and VPD-ICP/MS.

INTRODUCTION

Recently, trench isolation technology has been developed and applied to bipolar LSI production. Especially, ploy Si Chemical-Mechanical-Polishing (CMP) technique has made much improvement on deep trench isolation. [1]-[5] Fig.1 shows typical cross sectional transistor cell with poly Si trench isolation structure made by the CMP method. However, the wafer surface using a conventional CMP method is contaminated with silica particles and chemical impurities. The contamination produces pattern and crystal defects in the wafer surface layer after oxidation. It is difficult to remove them by the conventional cleaning technique. Therefore we studied a new post CMP cleaning method using special organic surfactant from the view point of contamination adhesion mechanism .

EXPERIMENT

The schematic diagram of the polisher, EBARA EPO-112, used in this study is shown in Fig.2. The polisher consists of polishing zone and wafer cleaning zone. The polishing zone is composed of wafer carrier and turntable. The wafer cleaning zone is composed of PVA brush and spin-dry stations [6]. The wafer is hold on side spin chuck and is rotated for cleaning. In the first stage, cleaning liquid is dropped onto a rotating PVA roll-like brush. In the second stage, the electric decomposed water (anode water) is dropped onto a wafer and spin dried. A wafer was polished on the turntable with a Rodeal IC1000/SUBA IV stacked polish pad and colloidal silica slurry of alkaline base and then , DI water was supplied on the turntable to polish the wafer only with water (We call this process water polish.). The films used in this study were PolySi, SiO2 (100 nm thick on Si). Remaining impurities on wafer after polishing and cleaning were evaluated by Vapor Phase Dissolution-Inductively Coupled Plasma/Mass Spectroscopy (VPD-ICP/MS) and TOF-SIMS(time of flight secondary ion mass spectrometry). Analyzed metallic elements were Fe, Cu and Cr and organic element was C. Number of particles were counted by a surface particle counter (SURFSCAN6200 and 6420 : Tencor instruments). Particles and their shapes analyzed EDX and AFM, respectively. Organic film thickness measured ellipse spectroscope. And transmittance of organic surfactant ware investigated by spectrophotometer.

RESULTS AND DISCUSSION

1. Conventional cleaning method

1.1 Metallic Contamination after CMP

We have first focused on metallic impurities on wafers after polishing. The concentrations of the metallic impurities in slurry used in our experiments are shown in Table 1. Fe was detected remarkably as metallic impurity. Many particles selectively located on the PolySi pattern region, because the wafer was only rinsed in DI water. Next, this wafer was oxidized as next process step. Many crystal defects were induced by metallic impurities. It is clear that crystal defects with 1E5 density produced dramatically in the wafer at the remaining Fe and Cu impurity contamination over 5E10 atoms/cm2. Therefor, it is necessary to keep the remaining metallic impurity below 1E10atoms/cm2 after polishing in order to establish the defect free wafer through the trench isolation process. [7]

1.2 Particles after CMP

There are many kinds of particles on the appeared film surfaces. The number of particles on Si wafer surface are strongly depend on the film termination after CMP process. Remaining impurities on the each wafer surface after polishing and cleaning were evaluated by EDX(Energy-dispersive X-ray) and AFM.

Fig .3 shows the residual particles on the two kinds of wafers after CMP. Over 15000 particles for each film still remained on the wafer (200 mm diameter)by DI water cleaning. We investigated the shape of particles in detail by AFM.

The AFM analysis on the Si revealed that most of the residual particles were pits, whose shapes were sub micron dimple like COP (Crystal Originated Particle). These pits were micro scratches introduced in wafer surface during CMP. Others were spherical type and disk type. In addition EDX analyzation revealed the elements of particles were composed of silicon, oxygen and organic. It is satisfactory to consider the particles as water mark. In general, disk type particles on Si, which are water marks, are easy to form on hydrophobic surface, but they are not on hydrophile surface. Therefore, we changed the polished Si hydrophobic surface to hydrophilic surface by the addition of organic surfactant. Then, the formation of water marks on the Si surface were prevented by the organic film formed on the Si surface.

The AFM analysis on the SiO2 film revealed that the 10%of particles were pit type. These particles were micro scratches same as Si wafer surface. Many micro scratches occurred on the SiO2 and Si surface induced some secondary particles in slurry during polishing. Particles were easily washed away with DI water.

2. New cleaning method

We developed a new cleaning method to remove the metallic impurities and the particles (scratch, water mark), which were not removed as discussed in section 1.2. In the case of water polishing process, contamination of the residual metallic impurity was reduced below 1E10 atoms/cm2 by the additions of organic surfactant in the DI water as shown in Fig.4. Furthermore, the number of particles on Si, SiO2, were decreased dramatically as shown in Fig.5. We paid attention to terminus elements on wafer surface, was considered as a reason why particles decrease if organic surfactant was dropped into water polishing. Fig.6 shows the relation between the organic film thickness of wafer surface with and without silica particles in organic surfactant for polySi CMP process and the cleaning method. The value of organic film thickness was under 1.0nm after PVA blush scrubbing without silica particle in the surfactant, and edge of wafer surface was hydrophobic. Therefore, many particles(water mark) remained on edge of wafer. On the other hand the film thickness was 4.0nm after PVA brush scrubbing with silica particle (0.1wt%). This wafer surface was hydrophilic perfectly. Therefore, surface of Si wafer after water polishing process were not contaminated by water mark.

We focused on the transmittance of the organic surfactant to appear the reason of this phenomenon. Fig.7 shows the relation of transmittance for the organic surfactant contain silica particle (0.1wt%) to pH. The transmittance of wafer surface decrease rapidly from pH 9.9 as the water polishing time increases in our experiment. The value of pH under 9.9 as the water polishing, an organic element of surfactant condensed as a silica particles were core on the wafer surface. Therefore solidify organic films formed on surface. In addition, many micro scratches on the Si, SiO2 surface were polished out during water polish using the organic surfactant with silica particles. But organic films on Si surface contain silica particles. In the second cleaning stage, the organic films removed out by anode water. Anode water have many OH radical, therefore it was be able to oxidize organic films on the wafer. Accordingly the organic films were replaced to oxide films. When anode water is supplied to the wafer surface, adsorbed of silica particles on the wafer will be remove out.

Fig.8 shows schematic drawing of new post-CMP cleaning process and characterization of surface for trench. Wafer after PolySi CMP process were contaminated by particles(Spherical type, micro scratches, water mark) and metallic impurities in the case of conventional cleaning method.(Fig.8-a) The new cleaning method were shown Fig.8-b and Fig.8-c. The first step, to prevent generating water marks, an organic film was formed on the wafer surface during water polishing. Many micro scratches on the wafer surface were polished in this cleaning step.(Fig.8-b) The second step, the organic films on the wafer surface were replaced to oxide film using anode water rinse and spin dried.(Fig.8-c) This new cleaning process was useful to CMP process and made the impurity level very low.

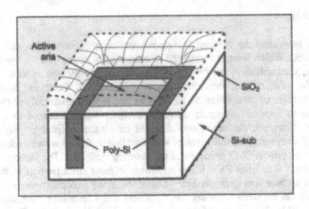

Fig.1 A schematic illustrating the transistor with trench isolation

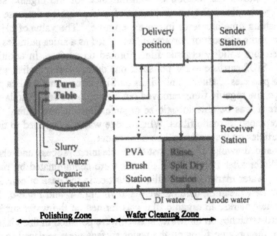

Fig.2 Schematic diagram of the CMP equipment (EPO-112)

Metallic impurities	Na	Fe	Cu	Cr
Concentration [ppm]	2370	40	<1	<1

Table.1 Metallic impurities in conventional slurry

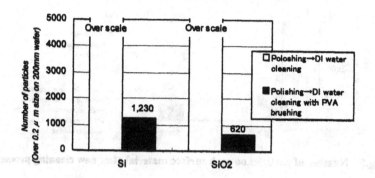

Fig.3 Number of particles on each surface material after conventional cleaning process

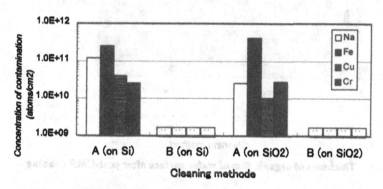

A : Polishing→DI water with PVA brushing
B : Polishing with surfactant→DI water with PVA brushing→Anode water cleaning

**Fig.4 Metal contamination on each kind of
material after conventional and new cleaning method**

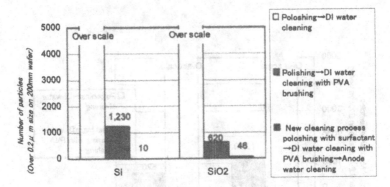

Fig.5　Number of particles on each surface material after new cleaning process

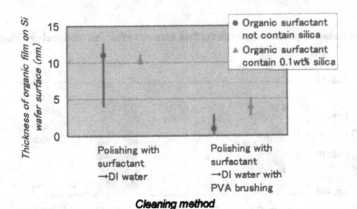

Fig.6　Thickness of organic film of wafer surface after post-CMP cleaning

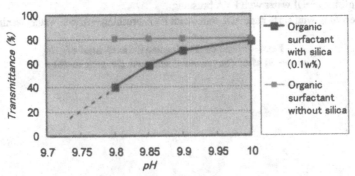

Fig.7　Transmittance of organic surfactant to pH

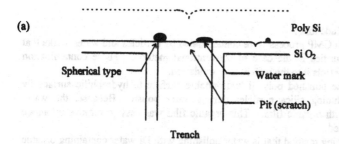

(a)

Poly Si

Spherical type

Si O₂

Water mark

Pit (scratch)

Trench

<Conventional cleaning method>

Water polishing	Cleaning	
Without organic surfactant	PVA	Spin dry
	DI Water	DI Water

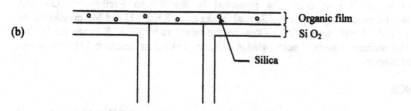

(b)

Organic film

Si O₂

Silica

Water polishing	Cleaning	
With organic surfactant	PVA	Spin dry
	DI Water	DI Water

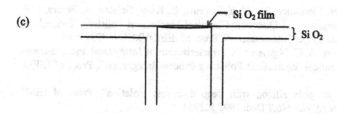

(c)

Si O₂ film

Si O₂

Water polishing	Cleaning	
With organic surfactant	PVA	Spin dry
	DI Water	Anode Water

<New cleaning method>

Fig.8 The process sequence of post CMP cleaning and characterization of surface for trench

CONCLUSIONS

From these results, we conclude that :
(1) The wafers after poly Si CMP process were contaminated by particles and water marks that including metallic impurities in the case of conventional method. These contamination introduced the crystal defects into the wafers after oxidation.
(2) It is easy to change the polished poly Si hydrophobic surface to hydrophilic surface by organic surfactant including silica particles during water polish. Because, the wafer surface was covered with organic film. This organic film was easy to remove by anode water rinse and spin dried.
(3) By using this new cleaning method that is water polishing with DI water containing organic surfactant and silica particles, the concentration of the remaining metal impurity on the wafer was reduced below 1E10 atoms/cm2 and caused no crystal defect after oxidation. Our new cleaning method is also applicable CMP planarization process for VLSI.

ACKNOWLEDGMENT

The authors would like to thank the personnel in the Silicon Facility at TOSHIBA TAMAGAWA for processing the wafers. Special thanks go to S.Kikuchi and M.Terasaki for the lithography, K.doi for the trench fill, Y.Otani for the trench etching, and K.Iwade for the line process. And authors gratefully acknowledge H.Kato, Tokuyama research Lab. for useful experimental support.

REFERENCE

[1] JM. Steigerwald, R. JGutmann, S. P. Murarka : "Electrochemical Effect in the Chemical Mechanical Polishing of Copper and Titanium Thin Film Used For Multilevel Interconnect Schemes", Proc. of IEEE-VMIC (1993) P.205

[2] S. Sivaram, H. Bath, R. Leggett, A. M aury, K.Monnig, R. Tolles : "Planarizing Interlevel Dielectrics by Chemical Mechanical Polishing", SOLD STATE TECHNOLOGY (1992.5) P.87

[3] S. Poon, A. R. Sitaram, B. Fiordalice, M. Woo, E. Prinz, C. King, Gelatos, A. Perera, D.B urnett, M. Hoffman : "Integration of Dielectric Chemical Mechanical Polishing Technology Advanced Circuits with Multilayer Interconnects", Proc. of IEEE-VMIC (1993) P.59

[4] W. Ong, S.Robles, S. Sohn, B. C. Nguyen : "Characterization of Inter-metal and Pre-metal Dielectric Oxides for Chemical Mechanical Polishing Process Integration", Proc. of IEEE-VMIC (1993) P.197

[5] S. A. Abbas : "Silicon on poly silicon with deep dielectric isolation", Proc. of IBM' Technical Disclosure Bulletin Vol. No.7 Dec. 1997 P.2754

[6] D. L. Hetherington : "The effects of double-sided scrubbing on removal of particles and metal contamination from chemical-mechanical-polished wafers", Proc. of DUMIC Conference 1995 ISMIC-101D/95/0156

[7] N. Miyashita, Y. Minami, I. Katakabe, J. Takayasu, M. Abe, T. Izumi : "Characterization of new post CMP cleaning method for trench isolation process" Proc. of Proceedings of 14th International Vacuum Congress. (1999)P.71

[8] E.Sirtl and A.Adler : Zeits.fur Metallk.,52 (1961) P.529

PCA CHARACTERIZATION OF RESIDUAL SUBSURFACE DAMAGE AFTER SILICON WAFER MIRROR POLISHING AND ITS REMOVAL

Y. OGITA*, K. KOBAYASHI*, H. DAIO**
*Kanagawa Institute of Technology, Atsugi, Kanagawa, 243-0292 Japan
**Showa Denko K. K., Chichibu, Saitama, 369-1813 Japan

ABSTRACT

Residual subsurface damages introduced by mirror polishing into Si CZ wafers degrade GOI in ULSI MOS devices. A removal of the damage throughout 9 times SC1 cleaning was systematically characterized as correlated between PCA signals measured by noncontact UV/mm-wave technique, GOI at 10MV/cm for MOS diodes with a thin gate-oxide thickness of 10nm, surface microroughness R_a measured by AFM. The same characterization was carried out for epitaxial wafers, as reference. Degraded GOI and PCA signal were recovered throughout 3 times SC1 cleaning and did not depend on R_a of 0.1-0.2 nm, which led to that the damage causes the degradation of GOI, but it can be removed by 3 times SC1 cleaning, and the damage depth was about 21nm. Further, the PCA signal well reflects to removal of the damage and degradation of GOI so that it can be a monitor for characterizing the removal and GOI. Direct observation of the damage using OSDA was carried out for the epitaxial wafer polished and SC1 cleaned. The OSDA indicated an image involving straight lines which disappeared after 3 times SC1 cleaning. This gave a direct evidence for catching up it by PCA and existence of residual subsurface damages.

INTRODUCTION

As the device dimension in ULSI reduces, thinner oxide of MOS devices is demanded. The thinner oxide is bringing a serious problem such as gate oxide breakdown which degrades GOI (gate oxide integrity). One of the cause has been known to come from residual subsurface damages induced by chemical-mechanical polishing (CMP) in the subsurface of a silicon wafer [1-3]. So, it is very important to investigate polishing condition to make the damage least and how to remove the damage before the device fabrication. At present, it is not so easy to detect the damage. The removal of the damage has been monitored by carrier lifetime measurements for hydrogen terminated Si wafers [4]. This result implied the carrier lifetime sensible to the damage. UV/ millimeter-wave PCD technique (Ultra Violet / millimeter-wave Photoconductivity Decay Technique) [5-7] has monitored removal of the damage [2,3,8], and revealed degradation of gate oxide integrity (GOI) due to the damage through the discussion correlating surface microroughness with GOI [2,3]. However, the relationship between PCA and surface microroughness was not necessarily clear, because PCA is also dependent on the surface microroughness.

In this study, the experiments have been carrier out as perfectly removing the question. PCA signal, GOI, and surface microroughness were measured removing the subsurface by repeating SC1 cleaning for as-polished wafers (PW) in commercial use and as-epitaxial wafers (EW) as reference without subsurface damages. GOIs were evaluated for MOS diodes with thinner 10 nm gate oxide as not including COP (crystal originated particle) as possible. Direct observation of the residual subsurface damages using OSDA (Optical shallow defect analyzer) was carried out for the epitaxial wafer polished and SC1 cleaned by the same process as it for PW samples.

PCA TECHNIQUE

The ultra violet laser-pulse having such a short pulsewidth T_w as 1ns and such a short wavelength as 337.1nm irradiates the silicon surface to excite excess carriers as shown in the bottom left figure of Fig. 1.The generated carriers are confined within the subsurface since extremely short penetration depth of 200Å. An amplitude (or intensity) of the photoconductivity

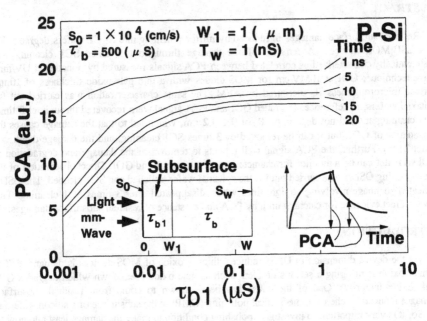

Fig. 1 Calculated PCA as a function of τ_{b1}. Two layer model for calculation (Bottom left). Definition of PCA (Bottom right).

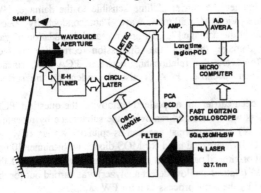

Fig. 2 UV/millimeter wave PCD maesurement system.

transient induced is defined as a photocnductivity amplitude (PCA) as shown in bottom right drawing in the figure.

Therefore, PCA detects carrier recombination interacting with imperfection within the subsurface. Using a two-layer wafer model consists of both subsurface and bulk layers as shown in inset of Fig. 1, calculated PCA signals are shown in Fig. 1 as a function of lifetime τ_{b1} within a subsurface depth w_1 of 1 μm which stands for imperfection in the subsurface, under parameters of bulk lifetime τ_b (500 μs) and surface recombination S_0 (10^4cm/s). PCA is responsible to τ_{b1} less than round of 0.1μs.

The measurement system employed to observe PCA signals is shown in Fig. 2. The N_2 laser with a pulse width of 1ns and a spot size of 3x5 mm^2 irradiated the surface of the silicon wafer sample to generate excess carriers. The sample placed apart from the aperture of the WG-10 waveguide by several millimeter was irradiated by the mm-wave of 100 GHz. PCA signals as photoconductivity variation detected with the diode from the reflected mm-wave were observed directly using the fast digitalizing oscilloscope (Lecroy, 6120) with 5G-sampling and 350 MHz - bandwidth.

EXPERIMENTAL

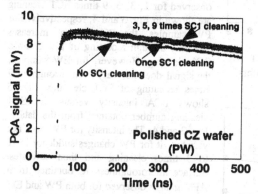

Fig. 3 PCA signals measured through 9 times SC1 cleaning of CZ wafers polished.

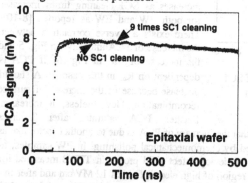

Fig. 4 PCA signals measured through 9 times SC1 cleaning of Epitaxial wafer.

Boron (B)-doped (100) plane p-type CZ-wafers (PW) with 10 Ωcm in resistivity and 150 mm in a diameter were chemomechanicaly polished in commercial use. A n-type (100) plane epitaxial wafer (EW) with 10 Ωcm and 150 mm was also prepared as the reference sample. Both wafers were treated with 9 times SC1 and SC2 cleaning. The respective cleaning was made for 6 minutes at 80°c using the SC1 solution with the composition of $NH_4OH:H_2O_2:H_2O=1:1:8$.

The removal depth was about 70 Å per once SC1 cleaning. The wafer samples with 1. 3. 5. 9 times SC1 cleaning through 9 times cleaning were employed for the experiments in this study. Gate oxide integrity (GOI) was evaluated for MOS diodes with a phosphorous doped polysilicon gate-electrode with an area of 16 mm^2, and an oxide thickness of 100 Å produced by the wet-oxygen oxidation at 850 °c. GOI was evaluated by TZDB (time zero dielectric breakdown) under the judge current of 1mA. To investigate correlation between PCA, the surface microroughness, and GOI, the microroughness R_a was observed for an area of 1 x 1 μm ^{2} by AFM (Seiko Instrument, SPA-360). PCA signals were measured using the UV/mm-wave PCD system mentioned above. In order to catch the direct evidence of existence of the residual damage induced

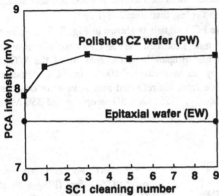

Fig. 5 PCA intensity as a function of SC1 cleaning number for both CZ and epi-wafers.

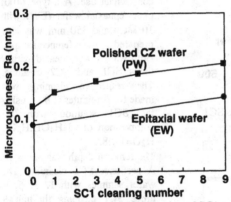

Fig. 6 Microroughness R_a as a function of SC1 cleaning number for both CZ and Epitaxial wafers.

by polishing, direct observation was made for following three kinds of p-type (100) plane epitaxial wafers with 10 Ωcm using OSDA (Hitachi): as-epitaxial wafer, as-polished wafer and as-3 times SC1 cleaned wafer prepared with the identical process for the PW samples.

RESULTS AND DISCUSSIONS

PCA signals for PW and EW samples observed for 1, 3, 5, 9 times SC1 cleaning are shown in Fig. 3 and 4, respectively. For PW samples, the PCA signal increases suddenly through repeating of 1 to 3 times SC1 cleaning. However, for EW samples, the signal does not vary even throughout 9 times repeating of SC1 cleaning. Fig. 5 shows PCA intensity versus the SC1 cleaning number obtained from the data in Fig. 3 and 4. The intensity for EW does not vary but it for PW changes suddenly for 1 to 3 times cleaning and then saturates. Surface microroughness R_a obtained from AFM images observed for both PW and EW is plotted as a function of the cleaning times as shown in Fig. 6. R_a continuously increases as SC1 cleaning times increases for both PW and EW as reported [8-10]. There exists a inverse correlation between the increase of PCA intensity in Fig. 5 and the increase of Ra in Fig. 6. If PCA is dependent on R_a, in this case, PCA should decrease because of the increase of surface recombination. Nevertheless, it increases. Further, PCA saturates after 3 times cleaning, but R_a does not so. These imply that the increase of PCA is due to another factor like as removal of the subsurface damages induced by chemomechanical polishing. In PW samples, as SC1 cleaning times increased to 3 times, the device defect yield profile in TZDB measured for MOS diodes shifted from 10 MV/cm to the region of high electric field of 12 MV/cm and after that the shift almost stopped. In EW samples, the profile existed at round of 11MV/cm did not shift throughout 9 times SC1 cleaning. Fig. 7 shows device defect density for the breakdown field over 10 MV/cm as a function of SC1 cleaning times for both PW and EW. The density D was estimated by D=-(1/A)lnY (Y: Yield of the breakdown, A: gate electrode area) [11]. The defect density in PW suddenly decreases through 3 times SC1 cleaning. On the other hand, it in EW does not vary throughout 9 times cleaning. The results in GOI and PCA should lead that the damage is caused by the mirror polishing and it can be removed by 3 times SC1 cleaning and its depth is about 21 nm, further, PCA well reflects the recovery of GOI and the behavior of removal of subsurface damage.

OSDA images observed for as-polished and as-SC1 cleaned EW samples are shown in Fig. 8. Many straight lines can be observed on the image as-polished, but disappeared in as-3 times cleaning. We think that the lines indicate a direct evidence of the damages induced by polishing and variation of the PCA signals.

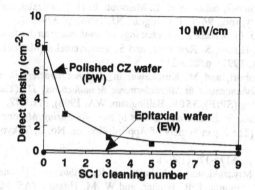

Fig. 7 Defect density as a function of clenaing number for both CZ and epitaxial wafers.

CONCLUSIONS

Both mirror polished CZ wafers in commercial use and epitaxial wafers which were subjected to repeatedly RCA (SC1+SC2) cleaning were systematically characterized as correlated between PCA, GOI, and surface microroughness. PCA signal intensity and GOI yield recovered through 3 times SC1 cleaning, while those for the epitaxial wafers did not varied even throughout 9 times cleaning. The results led that the damage causes the degradation of

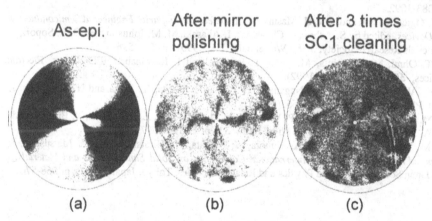

Fig. 8 OSDA images observed for as-epitaxial wafer (a), as-polished wafer (b), and as-3 times cleaned wafer with SC1 solution (c).

GOI, but it can be removed by 3 times SC1 cleaning, and the damage depth was about 21nm. The surface microroughness variation with SC1 cleaning times did not correlate to the measured PCA and GOI behaviors. PCA is concluded to well reflect to removal of the damage and GOI property, so that it can be a monitor for characterizing the removal and GOI. Observed OSDA images for the polished epitaxial wafer indicated straight lines which disappeared throughout 3 times SC1 cleaning. The lines should be to relate to the residual subsurface damage.and the direct evidence that PCA catches up the damage and the damage can be directly observed by OSDA.

REFERENCES

1. Y. Udo, M. Nagura, S. Samata, and H. Kubota, in *The Physics and Chemistry of SiO2 and the Si-Si O2 Interface-3*, edited by H. Z. Massoud, E. H. Poindexter, and C. R. Helms, (The Electrochemical Society, Inc., **96-1** Pennington, NJ, 1996), p. 379-387.
2. Y. Ogita, Y. Hosoda, and M. Miyazaki, in *Science and Technology of Semiconductor Surface Preparation*, edited by G. S. Higashi, M. Hirose, S. Raghavan, and S. Verhaverbeke, (Materials Research Society, **477**, Warrendale, PA, 1997), p. 209-214.
3. Y. Ogita, H. Shinohara, T. Sawanobori, and M. Kurokawa, in, *In-LineCharacterization Technique for Performance and Yield Enhancement in Microelectronic Manufacturing II(*, (The International Society for Optical Engineering (SPIE), **3509**, Bellingham, WA, 1998), p. 65-72.
4. M. Nakano, H. Masumura and H. Kudo in *Extended Abstracts of the 54th Spring Meeting*, The Japan Society of Applied Physics, (The Japan Society of Applied Physics, No. 2, Tokyo, 1993), p. 793(30a-ZD-6).
5. Y. Ogita, Semicond, Sci. Tech., **7**, 1, p. A175-179 (1992).
6. Y. Ogita, N. Tate, H. Masumura, M. Miyazaki, and K. Yakushiji, in *Recombination Lifetime Measurements i Silicon*, edited by D. C. Guputa, F R. Bacher, and W. M. Hughes, (ASTM, **STP 1340**, West Conshohocken, PA, 1998), p. 168-182.
7. Y. Ogita, K. Yakushiji and N. Tate, in *Semiconductor Silicon/1994*, edited by H. R. Huff, W. Bergholz, and K. Sumino, (The Electrochemical Society, Inc. **94-10**, , Pennington, NJ, 1994), p. 1083-1092.
8. Y. Ogita, M. Nakano, and H. Masumura, in *Defect and Impurity Engineered Semiconductors and Devices*, edited by S. Ashok, J. Chevsllier, I. Akasaki, M. M. Johnson, and B. L. Sopori, (Materials Research Society, **378**, Warrendale, PA, 1995), p. 591-596.
8 . T. Ohmi, M. Miyashita, M. Itano, T. Imaoka, and I. Kawanabe, IEEE Trans. Electron Devices, **39**, 3, p. 537-544 (1992).
9. K. Akiyama, N. Naito, M. Nagamori, H. Koya, E. Morita, K. Sassa and H. Suga, Jpn. J. Appl. Phys. Lett., **34**, p. L153-155 (1995).
10. M. Meuris, S. Verhaverbeke, P. W. Mertens, M. M. Heyns, L. Hellemans, Y. Bruynseraede and A. Philipossian, Jpn. J. Appl. Phys. Lett., **31**, p. L1514-1517 (1992).
11. M. Miyashita, H. Fukui, A. Kubota, S. Samata, H. Hiratsuka, and Y. Matsushita, in *Extended abstract of the 1991 International Conference on Solid State Devices and Materials*, (The Japan Society of Applied Physics and Related Societies, Tokyo, Japan, 1991), p. 568-570

USE OF DILUTE HF WITH CONTROLLED OXYGEN
FOR POST Cu CMP CLEANS

K. K. Christenson and C. Pizetti*

FSI International, 322 Lake Hazeltine Dr., Chaska, MN 55318, kchristenson@fsi-intl.com
*Metron Technologies Europe, christian3@compuserve.com

ABSTRACT

Wet cleaning of wafers during the semiconductor production process often requires uniform removal of a few nanometers of material. Ideally, a single cleaning chemistry can be found that etches all exposed features at a comparable rate. Etch rates near 1 nm/min are desired for batch process and near 10 nm/min for single-wafer processes. A mixture of 500:1 DHF (dilute HF) with dissolved oxygen controlled near parts-per-million (ppm) levels has been found to meet these requirements for post copper CMP (chemical-mechanical polishing) cleans with exposed SiO_2 and Cu metal.

INTRODUCTION

Figure 1 shows the cross section of a post Cu CMP device. Cleaning requires solvation of surface ionics along with the dissolution of approximately 5 nm of the surface to undercut and remove particles, remove surface damage and expose imbedded ions. Both DHF (dilute HF) and NH_4OH have been used as cleaning etchants in oxide CMP. Like the NH_4OH based SC-1, DHF is effective in removing particles from bulk SiO_2 and "bare" silicon with a native oxide [1]. DHF is preferred over NH_4OH because metal ions tend to be more soluble in acidic solutions. Also, the range of oxide etch rates available with DHF is considerably larger than those in NH_4OH solutions.

The etch rates of oxides in DHF vary dramatically with their chemical composition and deposition conditions. Ideally, it would be possible to control the oxide etch rate by the concentration of DHF, and the copper etch rate with a second, independent variable. Figure 2 shows the results of cleaning metallic copper deposited on hydrophobic silicon from a copper-spiked HF solution [2]. SC-2 solutions with varying HCl and H_2O_2 flows and a total flow of 1,760 ml/min were dispensed in the nitrogen-purged environment of a spray acid processor. The concentration of HCl had little effect on the total Cu removed as measured by TXRF. The concentration of H_2O_2 also had little effect, as long as some H_2O_2 was present. Even concentrated HCl solutions did not remove metallic copper in the nitrogen atmosphere without the addition of an oxidant. HCl solutions in an air ambient, however, do remove copper metal from wafers [3]. It may be possible to control the etch rate of copper in DHF solutions by varying the concentration of dissolved oxygen in the solution.

THEORY

Equations 1 to 3 are the individual chemical reactions believed to take place when etching copper in a DHF solution with dissolved oxygen. Figure 3 shows the solution pH as a

267

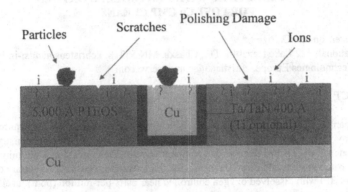

Figure 1: Cross section of a typical post Cu CMP device showing typical contamination and surface damage.

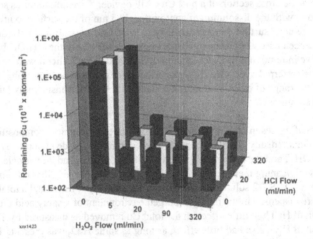

Figure 2: Removal of copper metal in SC-2 as a function of HCl and H_2O_2 concentrations as measured by TXRF.

function of concentration for HCl and HF. The pH of the HCl solutions in Figure 2 range from approximately −0.25 to 1.5. In this range, little effect of pH is seen on the total copper removed. Therefore, reaction (3), whose rate is expected to be proportional to $[H^+]^2$, is apparently not the rate-limiting reaction. While not investigated in this work, it is presumed that Reaction (3) is also not the rate-limiting reaction in the case of copper removal by DHF solutions near pH 2.

$$O_2(g) \rightarrow O_2(aq) \tag{1}$$
$$2Cu(s) + O_2 \rightarrow 2CuO(s \tag{2}$$
$$CuO(s) + 2H^+(aq) \rightarrow Cu^{++}(aq) + H_2O \tag{3}$$

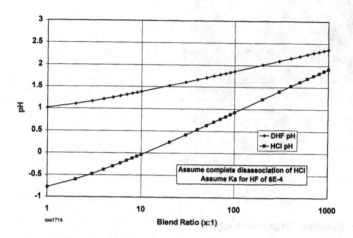

Figure 3: pH as a function for concentration for HCl and HF

The large increase in copper removal in Figure 2 between 0 and 5 ml/min H_2O_2 in the SC-2 mixture indicates the possibility of controlling the copper etch rate by controlling the oxidizing species. Figure 4 shows the Pourbaix diagram for copper [4]. Corrosion of copper in acidic aqueous solutions is *thermodynamically* favorable when the oxidation-reduction potential (ORP) of the solution E is greater than approximately 0.1 V.

The ORP of the acidic solution can be calculated using the Nernst equation applied to the chemical equilibria in equation 4 [5]. O_2 concentrations ranging from saturated to 1 ppb (a typical lower limit to O_2 concentration) give solution ORPs from 1.24 to 1.17 V. This range of ORPs and pH levels near 2 are marked by the small, dark rectangle centered above the word "corrosion" in the acidic portion of the Pourbaix diagram (Figure 4). This parameter space, used in this experiment, is entirely within the "corrosion" or etching regime. Therefore, etching of copper is *thermodynamically* favorable.

$$O_2 + 4H+ + 4e^- \leftrightarrow 2H_2O \qquad\qquad E^0 = 1.229 \text{ V} \tag{4}$$

The reaction rates for the reactions in equations 1 and 2 are expected to be linear with O_2 concentration. Given the diffusion coefficient D for O_2 in water of 2×10^{-5} cm^2/sec, it is possible to estimate the flux of O_2 to the surface [6]. Assuming 5 ppm of dissolved O_2 and a

269

10^{-3} cm boundary layer, the O_2 flux at the surface would be 10^{15} atoms/cm^2•sec. This flux is sufficient to support a 6-nm/min Cu etch rate.

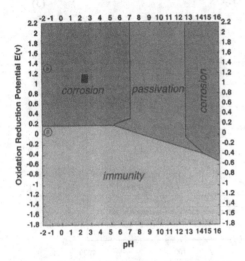

Figure 4: Pourbaix diagram for copper.

It is possible to determine whether the rate-limiting step is the diffusion of O_2 to the surface (Eq. 1) or the reaction of O_2 and Cu at the surface (Eq. 2) by measuring the variation in reaction rate with temperature. The rate of diffusion varies linearly with temperature, or 3% with a 10° C increase near ambient [6]. Chemical reaction rates typically grow exponentially with temperature and can double with each 10° C increase. The solubility of O_2 in water also varies with temperature.

EXPERIMENT

Etch experiments were carried out in a MERCURY® MP centrifugal spray acid processor [7]. In the acid processor, four cassettes of 25 wafers are mounted on a turntable in a sealed chamber. Cleaning chemistries and rinse water are atomized onto the wafers from sprayposts mounted at the center of the chamber (near the axis of rotation of the turntable) and on the outer chamber wall. When the DHF etchant is atomized, the tiny drops of liquid have a very high surface-area-to-volume ratio. The dissolved O_2 concentration in the DHF nearly equilibrates with the atomizing gas before striking the surface. The concentration of dissolved O_2 in the DHF can be controlled by the volume fraction of O_2 in the atomizing gas.

If very precise control of the dissolved O_2 is necessary, the DHF can be equilibrated with the O_2 in the atomizing gas in a commercial "degassification" module normally used to remove dissolved oxygen. If the O_2:N_2 blend in the degassification module, the atomizers and the chamber volume are matched, the O_2 concentration in the DHF will remain in equilibrium with the matched gas and then will remain constant throughout the system. Figure 5 shows the plumbing modifications to the spray processor for control of dissolved O_2.

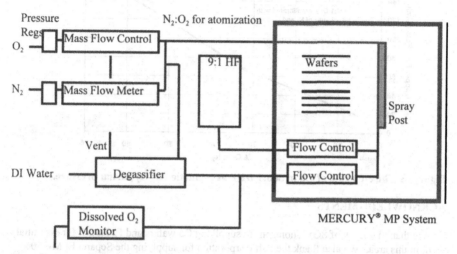

Figure 5: System to control precisely the dissolved gas concentration of liquids in a spray processor. Initial experimental work utilized rotometers in place of the mass flow controls to control the O_2:N_2 blend.

Using the hardware of Figure 5, 500:1 DHF was dispensed on 200 mm wafers for varying times. Oxygen concentrations in the matched gas varied from 0% to 65% by volume. The mass loss of Cu was determined either by weighing the wafer before and after each etch on an analytical balance or by directly measuring the thickness loss with x-ray fluorescence.

RESULTS

Figure 6 shows the etch rate of Cu in 500:1 DHF as a function of O_2 partial pressure. The etch rate is linear with O_2 concentration.

CONCLUSIONS

The etching of Cu metal in solutions of 500:1 DHF is linear with the concentration of dissolved O_2. Etch rates from 2 to 16 nm/min are controllable with O_2 partial pressures of 0.05 to 0.65 atm ($\sim$ 2.5 to 33 ppm O_2). The linear behavior indicates that the reaction is limited either by the transport of O_2 to the surface or the reaction of O_2 and Cu at the surface to form CuO. This DHF:O_2 chemistry shows great promise for post Cu CMP cleans.

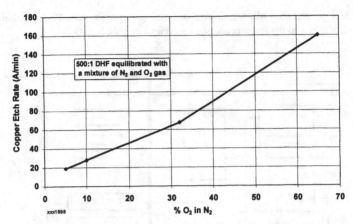

Figure 6: The etch rate of Cu in 500:1 DHF as a function of O_2 partial pressure.

ACKNOWLEDGMENTS

We thank D. Levy of SGS Thomson for supplying Cu wafers and F. Tardiff for his initial work in this area. We also thank the Pall Corporation for supplying the Separel EFM-350 degassification modules used in this work.

REFERENCES

1. K. Christenson et al., "Effects of Sequential Chemistries on Particle Removal," 4th International Symposium on Cleaning Technology in Semiconductor Device Manufacturing, R. Novak and J. Ruzyllo, eds., The Electrochemical Society, Pennington, NJ, PV95-20, 1996, pp. 567-574.
2. K. Christenson, S. Smith and D. Werho, "Removing Metallic Contaminants in an RCA-2 Clean as a Function of Blend Ratio and Temperature," *Microcontamination* magazine, Cannon Communications, June 1994.
3. K. Christenson, unpublished work.
4. M. Pourbaix, Atlas of Electrochemical Equilibria in Aqueous Solutions, 2nd ed., NACE International, Houston, 1974, p. 389.
5. D. Skoog and D. West, *Fundamentals of Analytical Chemistry*, 3rd ed., Holt, Rinehart and Winston, New York, 1976, p. 299.
6. J.S. Kirkaldy and D.J. Young, *Diffusion in the Condensed State*, Institute of Metals, London, 1987, pp. 94-95.
7. K. Christenson, "Benefits and Challenges of Centrifugal Spray Processor Technology," *Solid State Technology* magazine, PennWell, Tulsa, OK, 40, 12, Dec. 1998, p. 55.

SCANNING FORCE MICROSCOPE STUDIES OF DETACHMENT OF NANOMETER ADHERING PARTICULATES

J. T. DICKINSON, R. F. HARIADI, L. SCUDIERO, and S. C. LANGFORD
Department of Physics and Materials Science Program, Washington State University
Pullman, WA 99164-2814 jtd@wsu.edu

ABSTRACT

We employ salt particles deposited on soda lime glass substrates as a model system for particle detachment in chemically active environments. The chemical activity is provided by water vapor, and detachment is performed with the tip of a scanning force microscope. The later force required to detach nanometer-scale salt particles is a strong function of particle size and relative humidity. The peak lateral force at detachment divided by the nominal particle area yields an effective interfacial shear strength. The variation of shear strength with particle size and humidity is described in terms of chemically assisted crack growth along the salt-glass interface.

INTRODUCTION

The removal of submicron particles poses a severe challenge in a number of technologies, including optical components (mirrors and lenses) and high density integrated circuits. Whole technologies for particle removal have been developed, including laser assisted particle removal.[1,2] In integrated circuit manufacture, chemical mechanical polishing often serves this function, among others.[3] These processes typically employ both a liquid phase to reduce the adhesive forces and a mechanical stimulus to actually remove the particle. However, the mechanism of adhesive force reduction and the mechanical details of removal are not well understood.

In this work, we study the effect of humidity and applied shear stress on particle removal in a model particle/substrate system composed of single crystal NaCl grown onto soda lime glass substrates. The "chemical activity" of water vapor is readily varied over a wide range, making it an attractive choice for the chemically active phase. Both NaCl and soda lime glass are strongly hydrophilic, ensuring strong interactions with atmospheric water vapor. Further, the effect of humidity on crack growth in soda lime glass has been well studied (Ref. 4 and references therein). Stress is applied to the particles with the tip of a scanning force microscope (SFM), which may be compared to a well-characterized asperity in polishing processes. SFM tips can be instrumented to monitor the normal and lateral forces as the tip encounters the particle. Similar microscopies have been previously employed to study particle adhesion in a number of model systems, often involving metal particles on semiconductor substrates.[5-7]

In this work, we show that small amounts of water vapor dramatically lower the lateral force required to fracture the salt-glass bond as the SFM tip is drawn across the particle. We model this decrease in terms of the effect of water vapor on the interfacial surface energy. Particle size also affects the interfacial shear strength, presumably due to variations in the size of interfacial flaws relative to the total interface area.

EXPERIMENT

Submicron-sized NaCl crystals were deposited on soda lime glass substrates by dissolving a few grains (~ 1 mm³) of commercial salt in a drop of de-ionized water on a clean microscope slide. The solution was spread across the slide with a cotton swab and allowed to evaporate to dryness at ambient humidity and temperature. Both evaporation and sample storage were under ambient laboratory atmosphere conditions—typically 20-40% relative humidity (RH).

Particle observation and manipulation were performed on the stage of a Digital Instruments (Santa Barbara, CA) Nanoscope III scanning force microscope mounted in a controlled environment chamber. The humidity was adjusted by introducing a controlled mixture of dry and humidified air. The RH in the chamber was continuously monitored with a BioForce

273

Laboratory humidity sensor. Humidity variations in the course of an experiment were controlled to ± 1% absolute RH, with an estimated uncertainty of ± 2% absolute RH. This work employed triangular, 115-μm long, "wide" Si₃N₄ cantilevers from Digital Instruments of Santa Barbara, CA.

Particle detachment experiments were initiated by adjusting the RH in the experimental chamber and scanning the surface in the contact mode to find particles of the appropriate size. High scan rates (typically 42 μm/s) and low applied normal forces (5-20 nN) during this stage of the experiment were employed to minimize the effect of imaging on particle adhesion. When an appropriate particle was found, the tip was positioned for a linear scan that would intersect the NaCl particle at right angles, as close as possible to the center of an edge. At the beginning of the linear scan, the scan speed was reduced to 0.20 μm/s and the normal force increased (up to 320 nN). Particles of an appropriate size for a given humidity would detach from the substrate upon the tip's first encounter with the particle. Subsequent large area, low contact force scans over the region surrounding the initial particle position were made to determine the final position of the detached particle.

RESULTS

Images taken before and after particle detachment are shown in Fig. 1. A suitable particle was located in a high scan rate, low contact force image [Particle A in Fig. 1(a)]. After zooming in on the chosen particle, the scan rate was reduced. Scanning was continued until the tip was aligned with the center of the chosen particle [the white line in Fig. 1(b)]; then the vertical sweep was disabled and the contact force raised to a high value. After a single, high contact force scan, the contact force was lowered, vertical sweeping resumed, and the scan rate increased to normal values for imaging. Finally, large area scanning was resumed to search for the removed particle, marked A' in Fig. 1(b). At higher relative humidities, locating the removed particle can be difficult. Circumstantial evidence suggests that the absorbed water film can bind the removed particle to the SFM tip, which then drags the particle along without imaging it. Sometimes, the removed particle was found adhering to another particle in the image.

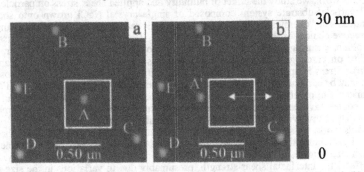

FIG. 1. Low contact force images of NaCl particles (a) before particle removal, and (b) after particle removal. The line in (b) shows the orientation of the linear scan used to remove Particle A, which was subsequently found at Position A'.

Typical lateral force signals during particle removal at 10% RH are shown in Fig. 2. The signal in Fig. 2(a) was acquired at a low applied normal forces (10 nN) and high scan rate (42 μm/s) prior to particle detachment. The lateral force as the tip passes over the salt particle is significantly lower than the lateral force as the tip passes along the glass. The reduction in lateral force ranges from 75-80% at 3% RH to 30% at 68% RH. The apparent coefficient of friction between the tip and the salt increases with increasing humidity, presumably due to the high affinity of water for NaCl; at high RH, the water layer may impede the progress of the SFM tip.

A typical lateral force signal during the linear scan that removes the particle is shown in Fig. 2(b). Immediately prior to the scan, the scan rate was decreased from 42 μm/s to 0.20 μm/s.

About a quarter of the way through the scan, the applied contact force was raised to 57 nN, producing the stepwise increase in lateral force. Several hundred milliseconds later, the tip encountered the particle; the height scan during particle encounter shows that the tip lifts slightly (~4 nm) up onto the edge of the particle. Detachment coincides with the peak in the lateral force signal. Because the tip is not in contact with the glass at the moment of detachment, the frictional force between the tip and the glass does not contribute to the lateral force at detachment. Thus the entire measured lateral force at detachment is applied to the particle, and it is this force that induces detachment.

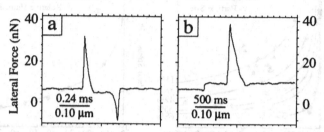

FIG. 2. Lateral force signals during (a) a low contact force scan while aligning the tip on the particle, and (b) the slow, high contact force scan used to remove the particle from the substrate. In (b), the stepwise increase in lateral force coincides with the increase in normal force, and the sharp peak corresponds to the completion of failure along the particle-substrate interface.

An upper bound on the energy per unit area required to fracture the interface is provided by the area under the rising portion of the lateral force vs. displacement plot prior to failure. In the case of Fig. 2(b), this amounts to about 0.05 J/m². In the fracture of more macroscopic, highly brittle samples (including both soda lime glass and NaCl), it is not uncommon for 2/3 of this energy to be dissipated via plastic deformation and similar processes. Since plastic deformation is strongly hindered in nanometer-scale systems, its role in the removal of nanometer-scale particles is an open question.

The lateral force required to remove a particle from the surface is a strong function of the scan rate. Low scan rates are considerably more effective than high scan rates in particle removal. Thus the duration of tip-particle contact, and not just the applied stress, determines whether the particle is removed. In particular, detachment is not merely due to the application of some critical stress greater than the intrinsic interfacial strength. We attribute failure under these conditions is to relatively slow, chemically enhanced growth of interfacial cracks at stresses below the ultimate interfacial strength. Detachment occurs when the crack velocity is high enough to progress along the entire interface during the brief duration of tip-particle contact, typically ~35 ms. Relative to other fracture processes chemically enhanced crack growth is typically quite slow (crack velocities < 10^{-4}-10^{-6} m/s). The very small particle sizes in this work allow for interfacial fracture on time scales of tens of milliseconds, even at these low crack speeds.

The nominal shear strength, σ_c (peak lateral force divided by the particle area, A) is plotted as a function of particle area for several relative humidities in Fig. 3(a). Note that the failure stress for 3% RH has been scaled (reduced) by a factor of ten for presentation purposes; the smallest particle at 3% RH failed at σ_c = 55 MPa! Raising the humidity from 3% to 11% dramatically lowers the interfacial shear strength. Particle size also affects shear strength, especially for the smaller particles. Both effects impose experimental limits on the size of particles amenable to study at the lowest humidities, where the SFM tip may break before a large particle will be removed. This is consistent with anecdotal reports of the difficulty of removing small particles under dry conditions. Importantly, increasing the humidity beyond 50% has little additional effect on the shear stress at failure. Similarly, the particle-size dependence becomes weak for particles with contact areas larger than about 3000 nm².

Figure 3(b) shows the nominal shear strength as a function of humidity for a set of particles with contact areas of ~5000 nm². The failure stress drops rapidly with increasing RH at the lower humidities, and falls more gradually at higher humidities. The dark line represents a least squares fit of the data to a model based on the expected dependence of interfacial binding energy on humidity, described below.

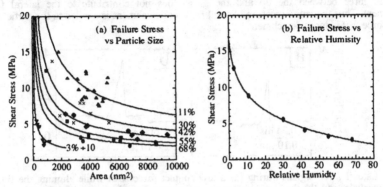

FIG. 3. Nominal shear stress at failure as a function of (a) particle/glass contact area at relative humidities ranging from 3% to 68%, and (b) relative humidity for a fixed particle size of ~5000 nm². The dark lines in (a) represent one-parameter, least squares fits to the data of the form $\sigma_c \sim A^{-1/2}$. The point for 3% RH in (b) has been extrapolated from the data, due to the difficulty of removing particles of this size at low humidities.

At high relative humidities, particle removal can be quite efficient. Figure 4 shows an attempt to "sweep" an 8×8 μm² area free of particles with a single scan. The resulting debris is piled up around the edges. This image illustrates the effectiveness of combined chemical and mechanical stimuli in particle removal.

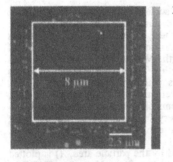

FIG. 4. SFM image of a region of a substrate cleared of adhering NaCl particles at 55% RH and a contact force of 51 nN.

DISCUSSION

To provide a framework for interpreting these results, we adopt several concepts from work on chemically enhanced crack growth. Although these chemical enhancements are not completely understood, studies by Wiederhorn, Michalske, Freiman, Bunker, Lawn, and others have advanced the field considerably. This work is reviewed in Lawn,[4] and for the purposes of discussion, we adopt the chemical approach outlined in his work. In this approach, crack growth is treated as a series of molecular-scale jumps of length a_0 at frequencies given by the product of an "attempt frequency" ν (a typical vibrational frequency) and a probability factor (the fraction of attempts yielding broken bonds). Ignoring the reverse process (crack healing), classical kinetic theory yields:

$$V = a_0 \, \nu \, \exp(-\Delta F/kT) \tag{1}$$

where V is the crack speed, k is Boltzmann's constant, T is the temperature, and ΔF is the free energy change associated with fracture in a chemically active environment.

The effect of stress on ΔF is conveniently described in terms of the stress intensity factor K, where ΔF is proportional to K. We note that this choice is somewhat controversial, in that ΔF is often taken to be proportional to the strain energy release rate, G, where $K = (G\,E)^{1/2}$ and E is Young's modulus. We adopt the stress intensity description because it provides a better description of our data. The difference between G and K is analogous to the difference between stress and volume activated processes during phase changes. In previous work on stress enhanced dissolution of single crystal surfaces, an analogous comparison of stress vs. volume activated mechanisms favored a volume activated process, consistent with the use of K here.[8,9]

In general, K is a complex function of sample geometry, including crack length, and is generally treated numerically. Nevertheless, K is expected to scale as

$$K \sim d\,\sigma_{xy}\,c^{1/2} \tag{2}$$

where σ_{xy} is the nominal shear stress applied to the particle (lateral force divided by particle/substrate contact area), c is the crack length, and d is a constant of order 3.[10] A complication in our loading scheme is the effect of the normal (compressive) stress exerted by the SFM tip on the adhering particles, σ_{yy}. This stress tends to close any existing crack and effectively reduces K. For the purpose of analysis, we assume this effect is independent of particle size.

Although we do not measure crack velocity per se, failure during the rather short tip-particle contact (typically 35 ms) requires a minimum crack velocity and thus a minimum K. The exponential dependence of crack velocity on σ_{xy} in Eq. (2) ensures that the majority of crack growth will occur when σ_{xy} is close to the peak shear stress, σ_c. For the purposes of analysis, we replace σ_{xy} with σ_c. Further, we assume that all particles of a given size (parameterized by A) have similar initial flaws, which we treat as initial "starter" cracks of length c in Eq. (2). Under these conditions, we expect σ_c to be proportional to $c^{-1/2}$.

Plots of σ_c as a function of the nominal particle/substrate contact area A are show in Fig. 3(a) for several values of relative humidity. The data were initially modeled the data with a two parameter curve fit ($\sigma_c \sim A^{-n}$) for each value of relative humidity. Each fit yielded $n = 0.5$ to within the uncertainty of the curve fitting procedure (typically ± 0.1). For simplicity, the curve fits in Fig. 3(a) represent one parameter least squares fits with $n = 0.5$, i.e., $\sigma_c \sim A^{-1/2}$. Since we expect $\sigma_c \sim c^{1/2}$, this dependence suggests that the initial flaw size is proportional to particle area ($c \sim A$). Interfacial flaws are expected to serve as starter cracks, and may be responsible for the particle size dependence of the shear stress required for detachment .

To model the effect of humidity on failure, we assume that the energy required for crack growth is lowered by an amount equal to the net decrease in surface energy due water sorption. Submonolayer coverages in many systems are described by the Langmuir isotherm, which gives $\Delta F = 2\Gamma_m \ln(p/p_0)$, where Γ_m is the adsorption energy per unit area for full (monolayer) coverage, and p is the partial pressure of the adsorbate in the surrounding atmosphere (proportional to relative humidity). Assuming that interfacial fracture occurs on experimental time scales occurs for cracks that reach a critical speed, V_c, and again replacing σ_{xy} with σ_c, the nominal shear stress at failure for particles of a given contact area will scale as:

$$\sigma_c = \gamma \ln\left[\delta/(1 + p/p_0)\right] \tag{3}$$

where γ, δ, and p_0 are fit parameters. A fit of Eq. (3) to experimental data for cubes with contact areas of 5000 nm^2 appears in Fig. 3(b). The model describes the data quite adequately.

Langmuir-style descriptions of adsorption at high relative humidities (several monolayers adsorbed) are somewhat strained. Nevertheless, they are probably reasonable for strongly hydrophilic substances like salt and glass. In these materials, the water/solid bonds are much stronger than water/water bonds, so that the total energy of adsorption is dominated by the adsorption energy of the first monolayer.

Several important avenues for future investigation present themselves. Experiments with better characterized substrates (e.g., single crystal silicon or noble metals) would allow for direct comparisons between experimental data an theoretical estimates of interfacial binding energies. Independent characterization of water adsorption isotherms would also facilitate detailed comparisons between theory and experiment. By changing the substrate, one can explore the effect of water vapor adsorption energy; this should strongly effect adhesion at low relative humidities, where substrate-vapor interactions are especially important. Equation (1) also suggests that particle removal can be a strong function of temperature. Measurements on variable temperature SFM stages would be of considerable interest.

CONCLUSIONS

The nominal shear stress required to remove particles from a soda lime glass substrate with an SFM tip is a strong function of particle size and relative humidity. At 3% RH, particles larger than about 500 nm on an edge could not be removed by the SFM tip at any accessible contact force. Increasing the RH to 30% dramatically reduces the stress required for particle removal and promotes the removal of much larger particles. Further increases in relative humidity yield smaller decreases in the failure stress. The humidity dependence of the shear stress can be described in terms of the change in interfacial energy due to water absorption.

The stress required for particle removal also scales roughly with the inverse square root of the particle area. The relevant fracture mechanics suggests that this is associated with an "initial flaw size" that is roughly proportional to their area.

Particle removal is an important factor in a number of technologies including integrated circuit manufacture and optical component manufacture. The small size of the relevant particles allows for strong chemical enhancements, despite the low crack velocities associated with chemical effects. *Combined* chemical and mechanical attack is most effective for particle removal.

ACKNOWLEDGMENTS

This work was supported by the National Science Foundation under Grant CMS-98-00230. We wish to thank Dr. John Hutchinson, Harvard University, for helpful discussions on fracture mechanics.

REFERENCES

1. A. C. Tam, W. P. Leung, W. Zapka, and W. Ziemlich, J. Appl. Phys. **71**, 3515-3523 (1992).
2. A. C. Tam, H. K. Park, and C. P. Grigoropoulos, Appl. Surf. Sci. **127-129**, 721-725 (1998).
3. G. Nanz and L. E. Camilletti, IEEE Trans. Semiconductor Manufacturing **8**, 382-389 (1995).
4. B. Lawn, *Fracture of brittle solids*, 2nd ed. (Cambridge University Press, Cambridge, 1993), Chapter 5.
5. E. Meyer, R. Luthi, L. Howald, and M. Bammerlin, in *Micro/Nanotribology and its Applications*, edited by B. Bhusan (Kluwer Academic, Dordrecht, 1997), pp. 193-215.
6. T. Junno, K. Deppert, L. Montelius, and L. Samuelson, Appl. Phys. Lett. **66**, 3627-3629 (1995).
7. C. Lebreton and Z. Z. Wang, J. Vac. Sci. Technol. B **14**, 1356-1359 (1996).
8. N.-S. Park, M.-W. Kim, S. C. Langford, and J. T. Dickinson, J. Appl. Phys. **80**, 2680-2686 (1996).
9. L. Scudiero, S. C. Langford, and J. T. Dickinson, Tribology Lett. **6**, 41-55 (1999).
10. J. W. Hutchinson, personal communication (1998).

AUTHOR INDEX

SUBJECT INDEX

Printed in the United States
By Bookmasters